LABORATORY MANUAL
FOR
ORGANIC
CHEMISTRY

Fourth Edition Revised

Kenneth F. Cerny • Marietta H. Schwartz
Christopher E. Katz

The authors wish to thank Gita Venkatakrishnan in the preparation of this volume and for her dedication to the instruction of organic laboratory techniques.

Contents

Instructor Information

Name:
Office:
email:

Organic Lab 1

My laboratory section is:　　　M　　　T　　　W　　　Th　　　F　　　AM/PM

My laboratory instructor is: _________________________ (email): _________________________

My laboratory TA is: _________________________ (email): _________________________

Organic Lab 2

My laboratory section is:　　　M　　　T　　　W　　　Th　　　F　　　AM/PM

My laboratory instructor is: _________________________ (email): _________________________

My laboratory TA is: _________________________ (email): _________________________

PART I

PRELIMINARY MATERIALS

SAFETY FIRST, LAST, AND ALWAYS

The organic chemistry laboratory is potentially one of the most dangerous of undergraduate laboratories. That is why you must have a set of safety guidelines. It is a very good idea to pay close attention to these rules, for one very good reason:

The penalties are only too real.

Disobeying safety rules is not at all like flouting many other rules. *You can get seriously hurt.* No appeal. No bargaining for another 12 points so you can get into medical school. Perhaps as a patient, but certainly not as a student. So, go ahead. Ignore these guidelines. But remember—

You have been warned!

1. *Wear your goggles.* Eye injuries are extremely serious but can be mitigated or eliminated if you keep your goggles on *at all times*. And I mean *over your eyes*, not on top of your head or around your neck. There are several types of eye protection available, some of it acceptable, some not, according to local, state, and federal laws. I like the clear plastic goggles that leave an unbroken red line on your face when you remove them. Sure, they fog up a bit, but the protection is superb. Also think about getting chemicals or chemical fumes trapped under your contact lenses before you wear them to lab. Then don't wear them to lab. Ever.

2. *Touch not thyself.* Not a biblical injunction, but a bit of advice. You may have just gotten chemicals on your hands in a concentration that is not noticeable, and, sure enough, up go the goggles for an eye wipe with the fingers. Enough said.

3. *There is no "away."* Getting rid of chemicals is a very big problem. You throw them from here, and they wind up poisoning someone else. Now there are some laws to stop that from happening. The rules were really designed for industrial waste, where there are hundreds of gallons of waste that have the same composition. In a semester of organic lab, there will be much smaller amounts of different materials. Waste containers could be provided for everything, but this is not practical. If you don't see the waste can you need, ask your instructor. When in doubt, *ask*.

4. *Bring a friend. You must never work alone.* If you have a serious accident and you are all by yourself, you might not be able to get help before you die. Don't work alone, and don't work at unauthorized times.

5. *Don't fool around.* Chemistry is serious business. Don't be careless or clown around in lab. You can hurt yourself or other people. You don't have to be somber about it- just serious.

6. *Drive defensively.* Work in the lab as if someone else were going to have an accident that might affect you. Keep the goggles on because *someone else* is going to point a loaded, boiling test tube at you. *Someone else* is going to spill hot, concentrated acid on your body. Get the idea?

7. *Eating, drinking, or smoking in lab.* Never. Without exception.

8. *The iceman stayeth, alone.* No food in the ice machine."It's in a plastic bag, and besides, nobody's spilled their product onto the ice yet." No products cooling in the ice machine, all ready to tip over, either. Use the scoop, and nothing but the scoop, to take ice cream out of the machine. And don't put the scoop in the machine for storage, either.

---

9. *Keep it clean.* Work neatly. You don't have to make a fetish out of it, but try to be *neat*. Clean up spills. Turn off burners or water or electrical equipment when you're through with them. Close all chemical containers after you use them. Don't leave a mess for someone else.

10. *Where it's at.* Learn the locations and proper use of the fire extinguishers, fire blankets, safety showers, and eyewash stations.

11. *Making the best-dressed list.* Keep yourself covered from the neck to the toes- no matter what the weather. That might include long-sleeved tops that also cover the midsection. Is that too uncomfortable for you? How about a chemical burn to accompany your belly button, or an oddly shaped scar on your arm in lieu of a tattoo? Pants that come down to the shoes and cover any exposed ankles are probably a good idea as well. No open-toed shoes, sandals, or canvas-covered footwear. No loose-fitting cuffs on the pants or the shirts. Nor are dresses appropriate for lab. Keep the midsection covered. Tie back that long hair. And a small investment in a lab coat can pay off, projecting that extra professional touch. It gives a lot of protection, too. Consider wearing disposable gloves. Clear polyethylene ones are inexpensive, but the smooth plastic is slippery, and there's a tendency for the seams to rip open when you least expect it. Latex examination gloves keep their grip and don't have seams, but they cost more. Gloves are not perfect protectors. Reagents like bromine can get through and cause severe burns. They'll buy you some time, though and can help mitigate or prevent severe burns. Oh, yes— laboratory aprons: they only cover the *front*, so your exposed legs are still at risk from behind.

12. *Hot under the collar.* Many times you'll be asked or told to heat something. Don't just automatically go for the Bunsen burner. That way lie *fire*. Usually-

No Flames!

Try a hot plate, try a heating mantle, but try to stay away from flames. Most of the fires I've had to put out started when some bozo decided to heat some flammable solvent in an open beaker. Sure, there are times when you'll have to use a flame, but use it away from all flammables and in a hood (Figure 1), and only with the permission of your instructor.

13. *Work in the hood.* A hood is a specially constructed workplace that has, at the least, a powered vent to suck noxious fumes outside. There's also a safety glass or plastic panel you slide down as protection from exploding apparatus

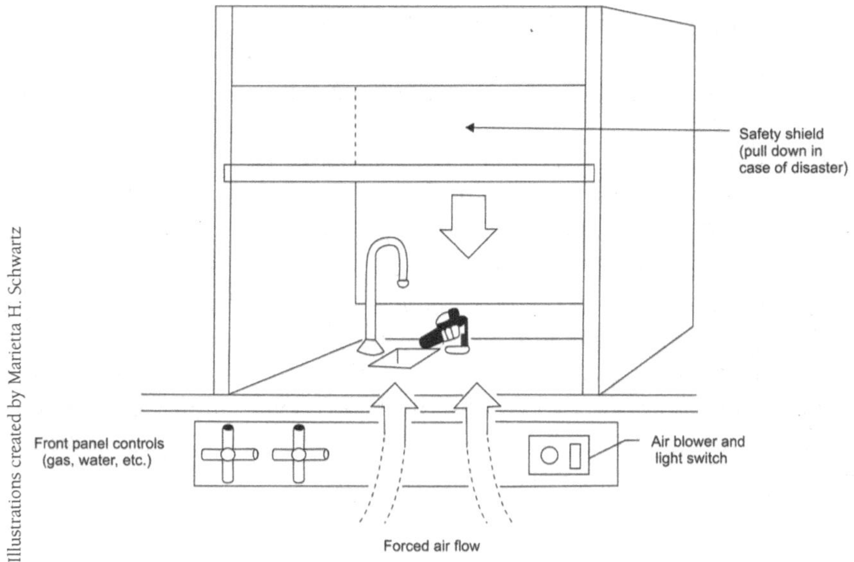

Figure 1—*A typical hood.*

(Figure 1) if it is all possible, treat every chemical (even solids) as if toxic or bad-smelling fumes can come from it, and carry out as many of the operations in the organic lab as you can *inside a hood*, unless told otherwise.

14. ***Keep your fingers to yourself.*** Ever practiced "finger chemistry"? You're underprepared so you have a lab book out, and your finger points to the start of a sentence. You move your finger to the end of the first line and do that operation—

"Add this solution to the beaker containing the ice-water mixture"

And WHOOSH! Clouds of smoke. What happened? The next line reads—

"very carefully as the reaction is highly exothermic."

But you didn't read that line, or the next, or the next. So you are a danger to yourself and everyone else. Read and take notes on any experiment before you come to the lab.

15. ***What you don't know can hurt you.*** If you are not sure about an operation, or you have any question about handling anything, *please* ask your instructor before you go on. Get rid of the notion that asking questions will make you look foolish. Following this safety rule may be the most difficult of all. Grow up. Be responsible for yourself and your own education.

16. ***Blue Cross or Blue Shield?*** Find out how you can get medical help if you need it. Sometimes during a summer session, the school infirmary is closed, and you would have to be transported to the nearest hospital.

17. ***Never carry chemicals to your instructor to show them.*** You're preparing a compound, and you have a question about what to do next. Perhaps your instructor is in the instrument room, or getting materials in the stock room, or even just at the next bench with another student. Don't carry your intermediate products to your instructor and ask that she come over and see what you are talking about. Do not ever carry this stuff out of the main lab, or across or down a hallway—ever.

These are a few of the safety guidelines for an organic chemistry laboratory. You may have others particular to your own situation.

ACCIDENTS WILL NOT HAPPEN

That's an attitude you might hold while working in the laboratory. You are not going to do anything or get anything done to you that will require medical attention. If you do get cut, and the cut is not serious, wash the area with water. If there's serious bleeding, apply direct pressure with a clean, preferably sterile, dressing. For a minor burn, let cold water run over the burned area. For chemical burns to the eyes or skin, flush the area with lots of water. In every case, get to a physician if at all possible.

If you have an accident, *tell your instructor immediately. Get help!* This is no time worry about your grade in lab. If you put grades ahead of your personal safety, be sure to see a psychiatrist after the internist finishes.

DISPOSING OF WASTE

Once you do our reaction, since your mother probably doesn't take organic lab with you, you'll have to clean up after yourself. I hesitated to write this section for a very long time because the rules for cleaning up vary greatly according to, but not limited to federal, state, and local laws, as well as individual practices at individual colleges. There are even differences—legally—if you or your instructor do the cleaning up. And, as always, things do seem to run to money—the more money you have to spend, the more you can throw away. So there's not much point even trying to be authoritative about waste disposal in this little manual, my classification scheme may not be the same as the one you'll be using. When in doubt, **ask! Don't just throw everything into the sink. Think.**

Note to the picky: The word nonhazardous, as applied here, means relatively benign, as far as organic laboratory chemicals go. After all, even pure water, carelessly handled, can kill you.

1. *Nonhazardous insoluble waste.* Corks, sand, alumina, silica gel, sodium sulfate, magnesium sulfate, and so on can probably go into the ordinary wastebaskets in the lab. Unfortunately, these things can be contaminated with hazardous waste (see following discussion), and then they need special handling.
2. *Nonhazardous soluble solid waste.* Some organics, such as benzoic acid, are relatively benign and can be dissolved with a lot of tap water and flushed down the drains. But if the solid is that benign, it might just as well go out with the nonhazardous insoluble solid waste, no? Check with your instructor; watch out for contamination with more hazardous materials.
3. *Nonhazardous soluble liquid waste.* Plain water can go down the drains, as well as water-soluble substances not otherwise covered below. Ethanol can probably be sent down the drains, but butane? It's not that water soluble, so it probably should go into the general organic waste container. Check with your instructor; watch out for contamination with more hazardous materials.
4. *Nonhazardous insoluble liquid waste.* Compounds such as 1-butanol (previously discussed), diethyl ether, and most other solvents and compounds not covered otherwise. In short, this is the traditional "organic waste" category.
5. *Generic hazardous waste.* Pretty much all else not listed separately. Hydrocarbon solvents (hexane, toluene), amines (aniline, triethylamine), amides, esters, acid chlorides, and on and on. Again, traditional "organic waste." Watch out for incompatibilities, though, before you throw just anything in any waste bucket. If the first substance in the waste bucket was acetyl chloride and the second is diethylamide (both hazardous liquid wastes), the reaction may be quite spectacular. You may have to use separate hazardous waste containers for these special circumstances.
6. *Halogenated organic compounds.* 1-Bromobutane and *tert-* butyl chloride, undergraduate laboratory favorites, should go into their own waste container as "halogenated hydrocarbons." There's a lot of agreement on this procedure for these simple compounds. But what about your organic unknown, 4-bromobenzoic acid? I'd have you put it and any other organic with a halogen in the "halogenated hydrocarbon" container and not flush it down the drain as a harmless organic acid, as you might do with benzoic acid.
7. *Strong inorganic acids and bases.* Neutralize them, dilute them, and flush them down the sink. At least as of this writing.
8. *Oxidizing and reducing agents.* Reduce the oxidants and oxidize the reductants before disposal. Be careful! Such reactions can be highly exothermic. Check with your instructor before proceeding.
9. *Toxic heavy metals.* Convert to a more benign form, minimize the bulk, and put in a separate container. If you do a chromic acid oxidation, you might reduce the more hazardous Cr^{6+} to Cr^{3+} in solution and then precipitate the Cr^{3+} as the hydroxide, making lots of expensive-to-dispose-of chromium solution into a tiny amount of solid precipitate. There are some gray areas, though. Solid manganese dioxide waste from a permanganate oxidation should probably be considered a hazardous waste. It can be converted to a soluble Mn^{2+} form, but should Mn^{2+} go down the sewer system? I don't know the effect of Mn^{2+} (if any) on the environment. But do we want it out there?

   LABORATORY MANUAL FOR ORGANIC CHEMISTRY

MIXED WASTE

Mixed waste has its own special problems and raises even more questions. Here are some examples :

1. *Preparation of acetaminophen (Tylenol):* a multistep synthesis. You've just recrystallized 4-nitroaniline on the way to acetaminophen, and washed and collected the product on your Büchner funnel. So you have about 30–40 mL of this really orange solution of 4-nitroaniline and by-products. The 4-nitroaniline is very highly colored, the by-products probably more, so there isn't really a lot of solid organic waste in this solution, not more than perhaps a hundred milligrams or so. Does this go down the sink, or is it treated as organic waste? Remember, you have to package, label, and transport to a secure disposal facility what amounts to 99.9% perfectly safe water. Check with your instructor.

2. *Preparation of 1-bromobutane.* You've just finished the experiment and you're going to clean out your distillation apparatus. There is a residue of 1-bromobutane coating the three-way adapter, thermometer, inside of the condenser, and the adapter at the end. Do you wash the equipment in the sink and let this minuscule amount of a halogenated hydrocarbon go down the drain? Or do you rinse everything with a little acetone into yet another beaker and pour that residue into the "halogenated hydrocarbon" bucket, fully aware that most of the liquid is acetone and doesn't need special halide treatment? Check with your instructor.

3. *The isolation and purification of caffeine.* You've dried an ethylene chloride extract of caffeine and are left with ethylene chloride-saturated drying agent. Normally a nonhazardous solid waste, no? Yes. But where do you put it while the ethylene chloride is on it? Some would have you put it in a bucket in a hood and let the ethylene chloride evaporate into the atmosphere. Then the drying agent is nonhazardous solid waste. But you've merely transferred the problem somewhere else. Why not just put the whole mess in with the "halogenated hydrocarbons"? Usually, "halogenated hydrocarbons" go to a special incinerator equipped with traps to remove HCl or HBr produced by burning. Drying agents don't burn very well, and the cost of shipping the drying agent part of this waste is very high. What should you do? Again, ask your instructor.

In these cases, as in many other questionable situations, I tend to err on the side of caution and consider that the bulk of the waste has the attributes of its most hazardous component. This is, unfortunately, the most expensive way to look at the matter. In the absence of guidelines,

1. Don't make a lot of waste in the first place.
2. Make it as benign as possible. (Remember, though, such reactions can be highly exothermic, so proceed with caution.)
3. Reduce the volume as much as possible.

Oh. Try to remember that sink drains can be tied together, and if you pour a sodium sulfide solution down one sink while someone else is diluting an acid in another sink, toxic, gagging, rotten-egg-smelling hydrogen sulfide can back up the drains in your entire lab, and maybe even the building.

LABORATORY SAFETY ACKNOWLEDGMENT

Organic Lab 1

NAME ____

STUDENT ID ____

LAB SECTION (CIRCLE ONE):   MON   TUES   WED   THUR   FRI   AM/PM

- I will wear safety goggles whenever *anyone* in the laboratory is doing an experiment.
- I will not eat, drink, or smoke while in the laboratory.
- I know the location of the emergency eye wash and shower.
- I will pour nothing down the sink without explicit instruction from the laboratory instructor.
- I will report any injury or unsafe condition immediately to the laboratory instructor.

SIGNATURE ____

DATE ____

Laboratory Safety Acknowledgment

Organic Lab 2

NAME ____

STUDENT ID ____

Lab Section (Circle One):   Mon   Tues   Wed   Thur   Fri   AM/PM

- I will wear safety goggles whenever *anyone* in the laboratory is doing an experiment.
- I will not eat, drink, or smoke while in the laboratory.
- I know the location of the emergency eye wash and shower.
- I will pour nothing down the sink without explicit instruction from the laboratory instructor.
- I will report any injury or unsafe condition immediately to the laboratory instructor.

SIGNATURE ____

DATE ____

BEST PRACTICES, ADVICE, SUGGESTIONS

1. GENERAL COMMENTS

The following items are required for your organic chemistry laboratory experience:

1. This lab text, of course
2. *Student Lab Notebook: 50 Carbonless Duplicate Sets* , Hayden-McNeil Specialty Products
3. Safety goggles. (these will be provided at check-in)

This syllabus has been prepared in order to make your laboratory experience as smooth, enjoyable, and invigorating as possible. It will be assumed that you have read and understood its contents before the first laboratory period.

There has been more than sufficient time allotted for the completion of each experiment. Thus, you should have no problem in completing all the assigned experiments. Attendance will be taken at every scheduled lab, and only under very special circumstances will students be allowed to "catch up" in a different lab section; therefore, absence from your scheduled lab section *should be avoided.* Laboratory sessions, including the prelab instructions, will begin on time and will end at the scheduled time. Lateness is NOT encouraged, and forcing the instructor to remain past the scheduled ending time will unquestionably affect your grade undesirably.

The laboratories, glassware, chemicals, and instruments are ours; take pride in cleanliness, neatness, and economy. It is our University; please work in such a fashion as to make the best and most efficient use of our facilities. Please leave the hoods and balance areas tidy, and put your equipment and glassware away neatly and carefully.

2. GRADING

Your laboratory grade will be based on the following criteria:

*   Laboratory reports
*   Prelabs
*   Final quiz
*   Yield and quality of products
*   Instructor's evaluation of technique and general preparedness

3. THE LABORATORY NOTEBOOK

Your laboratory notebook will be the primary record of your experimental work. The required notebook is the *Student Lab Notebook: 50 Carbonless Duplicate Sets)* from Hayden-McNeil Specialty Products, which is available in the bookstore (a copy also came bundled with this laboratory manual). No substitutes will be accepted. All your entries must be made in **ink**. Your notebook must begin with a Table of Contents. The inside cover of this notebook has one already set up for you; simply fill this in as you progress. The pages are already numbered sequentially. Science requires absolute honesty, and no deliberate misrepresentation of an experiment will be tolerated; fabrication of data will result in an F grade for the course!

Records of your experimental work must answer these questions: **When** did you do the work? **What** are you trying to accomplish with this experiment? **How** did you do the experiment? **What** did you observe? **How** do you explain your observations and results? Your notebook must be written with accuracy and completeness. It must be organized and legible, BUT it does not need to be a "work of art." Write all data (weights, volumes, physical constants, etc.) and pertinent information directly in your notebook. Using scraps of loose paper for recording data is unacceptable, as they may easily be lost.

Your experiments will be one of two types: a *technique* experiment or a *synthesis* experiment. Each new experiment must start on a new page and the date entered each day you work in the lab on that experiment. If you have an unfinished experiment when you start the next one, leave a page blank for completion. If you must continue on a noncontiguous page, include "continued on page" and "continued from page" references. Your laboratory notebook will provide copies, and as you work in the lab, be sure that you are in fact making copies of your work. You will turn these copies in as part of your lab report for each experiment. Each experimental record needs to be written in three steps: prelab, during lab, and postlab. It must contain the following items for each experiment performed.

BEFORE You Come to the Laboratory

These things will help prepare you for the experiment:

Title: Use a title that clearly shows what you are doing in this experiment. This often answers the question, what are you trying to accomplish?

Date: Use the date that you actually begin preparing for the experiment. Also date each actual working lab day.

Reference: Give the reference for this particular experiment. It may be as simple as Cerny/ Schwartz/ Katz, pp. 55–58.

Balanced Chemical Reaction: Write a balanced chemical reaction (using both structures and names) showing the overall process for any *synthetic* experiment (a *technique* experiment will not have an equation). If your experiment involves a separation, a flowchart tracking the location of the various compounds and solvents is appropriate as well.

Table: Include all reactants and products and organic solvents. The table normally lists molecular weights, the number of moles and grams of reactants used, densities of liquids used, boiling points or melting points (at *room temperature*) for reagents used, and solubility data if applicable. Sometimes a refractive index is necessary. For information about how to find these physical constants, see *A Guide to the Chemical Literature*.

Outline of Procedure: Write an outline of the experimental steps in sufficient detail so that the experiment could be performed without reference to this lab textbook. Do not copy the text verbatim; and certainly do not paste in photocopies of the lab text—you must write out your own procedure outline. Students will not be allowed to use the lab text in the laboratory; therefore, proper preparation is essential.

DURING the Laboratory Session(s)

You should first write the date; this will reflect when you actually started (or continued) the experiment. Then note in **ink** (please use every other line for clarity) the actual procedure that you followed and any observations that you noticed. Your descriptions should be such that someone else could repeat your experiment using only this procedure from your notebook. YES: This indeed does mean that you essentially may rewrite the outline you previously wrote before coming into the lab, but this time include **what you really did** with your observations. In brief, write down **what you see and what you do as you see it and as you do it**. This is extremely important! Because organic chemistry is an experimental science, your observations are crucial to your success. Things that may seem insignificant may be important in understanding and explaining your results later. Some typical observations might include the following:

*   The solution turned cloudy upon cooling.
*   The solution turned a bright pink color.
*   4 mL more solvent was needed to dissolve the solid.
*   A yellow precipitate appeared.
*   The white precipitate dissolved upon addition of the acid.
*   A puff of white smoke appeared.
*   An infrared spectrum was run on the liquid (neat).

Do not confuse the observations with their interpretation. For example, if you did a test for halides, "positive" is an interpretive conclusion, but "yellow precipitate formed" is an observation. **Both** the observation and the conclusion should be recorded in your notebook. The detail recorded for a particular technique will vary throughout the course as your knowledge and skills improve. For example, the first time you do a recrystallization, you may take two or three pages to detail all the steps involved. Later, the whole process may be summarized: "The crude product was recrystallized from 95% ethanol to give …"

AFTER the Experimental Work Is Completed

After the experiment, evaluate and interpret your results. Include calculation of the percent recovery or percent yield, interpretation of physical and spectral data, a conclusion summary, and answers to any postlab questions.

Percent Recovery: Whenever you separate and purify a given reagent, it was already "made" (that is, you did not make it). So a percent recovery is appropriate:

$$\text{Percent Recovery} = \frac{\text{weight of isolated product}}{\text{weight of starting mixture}} \times 100\%$$

Percent Yield: Write out your calculations for the percent of the theoretical yield that you actually obtained; remember to base your calculation on the limiting reactant:

$$\text{Percent Yield} = \frac{\text{amount of product actually produced}}{\text{theoretical yield of product}} \times 100\%$$

Conclusion/Summary: Include a succinct discussion of your results, drawing conclusions consistent with your observations and physical data. These remarks must include a thorough interpretation of your IR spectra (only if you actually took the spectrum) and any other analytical results. Properly labeled spectra must be firmly attached to your report.

4. YOUR REPORT FOR EACH EXPERIMENT

Roughly two weeks after you are scheduled to complete each experiment (in the Fall-Spring course sequence) you will be required to turn in the following:

1. **Title page** (no copy is necessary; typed on a separate sheet of paper): This gives the title of the experiment, your name, the date, the lab section, and the semester.
2. **Abstract** (no copy is necessary; typed on a separate sheet of paper): Abstract is defined by Webster's *New World Dictionary* as "that which presents the substance or general idea in brief form; concise; condensed." Write an abstract of the whole experiment; 2–4 sentences that describe the purpose, procedures, and results (**This does NOT go in your notebook**). Use your best writing skills: you will be graded on conciseness and accuracy. The purpose is to guide the reader to the important parts of the full report.
3. **COPY pages**: The copy pages of your lab notebook with appropriate IR spectrum if obtained by you. Note that you must start each new experiment on a new page to be able to turn in the notes one experiment at a time.
4. **Postlab Questions** Occasionally there are questions to be answered after the experiment. These are written in the lab text *after* the experiment. Typed answers to these questions should be included at the end of your report on a separate sheet. (**This does NOT go in your notebook**).
5. **Product** (if applicable): For most experiments, you turn in your product for grading. The products will be graded for purity and yield. Products must be identified correctly!

Items #1-4 should be stapled together in that order. These will be graded and returned to you in one to two weeks. NOTE: Even if you are doing more than one experiment in a given lab section, each experiment is turned in as its own separate lab report. (This is why each page in your lab notebook must be about only one (1) experiment!)

5. PRELAB PREPARATION AND LABORATORY TECHNIQUES

Students will be evaluated by their instructor according to their degree of preparedness and knowledge of the experiment. Safety is of the utmost importance; safety goggles or glasses (OSHA approved, with side shields; *not* plain prescription glasses) are to be worn at all times. Contact lenses are not recommended in the organic labs, as organic solvent vapors tend to have a detrimental effect on the lifetime of the lenses. Instructors have the authority to ask a student to leave the lab for the day, after appropriate warning. The instructors will be evaluating the students on questions that are asked by the students (should students know the answer if they are well prepared?) and on questions that are asked by the instructors. Students may be asked to leave if they are not properly prepared. Student performance in chemical manipulations will also be considered. Some of the many undesirable actions are leaving reagent bottles open; leaving the hood, balance area, or bench dirty; endangering self and/or others; careless handling of equipment; waste of chemicals; and improper and/or hazardous disposal of waste. Performance of assigned cleanup chores after each laboratory session is *not* optional.

6. PRELABS AND QUIZ

Prelab assignments are due at the beginning of the lab period. Failure to complete the prelab before that time will result in significant loss of points for this portion of your grade. There will be a summary quiz given at the end of the semester (see your laboratory schedule for specific dates). This quiz is open laboratory notebook. This will be similar to the individual prelabs but broader in scope, as it will cover all experiments performed during the semester.

7. MAKE-UP LAB

If, for *good* and *valid* reasons (to be determined by the instructor), you have to miss a lab session, you must obtain permission for a makeup lab from both your own lab instructor and the instructor whose section you wish to attend. A permission slip will be given to you by your lab instructor, and this slip must be presented to the instructor in whose section the lab is being made up. Makeup lab time (other than that scheduled at the end of the semester) is not guaranteed—try not to need it!

When you work in a lab section other than your own, be courteous. You do not have the right to make up a lab; it is a privilege. The students in that lab section have priority use of all space, equipment, and chemicals. Please do not abuse this privilege.

8. CHECK-IN AND CHECK-OUT

At check-in you will use the sheet (found in this laboratory manual) listing all of the apparatus and glassware that should be present in your assigned drawers. Make sure all the glassware and other equipment is present and in good condition. Specific procedures will be explained by your laboratory instructor. The check-in sheet, along with the combination card, must be turned in to your laboratory instructor at the end of the check-in period.

At check-out you will be given the same sheet by your instructor. Repeat the check-in process, making sure that you leave your drawer in as good a condition as you found it (or better). After you have checked out, you may determine melting points, refractive indexes, weights, etc. However, no "wet chemistry" (actual performance of experiments or completion of experiments) will be allowed. ***Students who do not check out of the lab WILL fail the course***.

9. SAMPLE LABEL FORMAT

A model label for the sample vials is shown below. These labels should be taped over, as they may not stick well to the sample vials. (Tape is available in the laboratory.)

<table>
<tr><td>

Fluorenone
3.24 g (60%)
mp. 80–82 oC

</td><td>

May 5, 2017

J .C. Smith
Organic Lab 1, Wed. lab, Spring 17

</td></tr>
</table>

10. MOST COMMON MISTAKES IN LABORATORY NOTEBOOKS

Format Problems:

1. Missing dates and titles of experiments.

2. Missing reactions and calculations (yield, etc.).

3. Missing tables of reagents (or some of the necessary data).

4. Missing necessary graphs (should be plotted in the notebook or on graph paper that is *permanently* attached to the notebook—no loose sheets!).

Table of Contents:

Must be present—refers to experiments performed and the page(s) on which those experiments appear in the notebook.

Organization:

The notebook should be a chronological record of what was done in the laboratory, with the appropriate calculations and conclusions added immediately after the experiment. DO NOT leave blank pages. If you run out of space, simply write "continued on page X" at the bottom of the page, and continue on the next available page in the notebook.

EACH EXPERIMENT SHOULD BEGIN WITH THE FOLLOWING (PREPARED AHEAD OF TIME)!

1. Title of the experiment

2. Reference

3. Reaction or separation being performed (write out the chemical notation)

4. Table of physical constants (look up BEFORE lab!)

5. The basic procedure(s) for the experiment(s) being performed in lab, in enough detail that reference to the lab manual is unnecessary.

DURING THE COURSE OF THE LAB, THE FOLLOWING INFORMATION SHOULD BE WRITTEN:

The procedure for the experiment(s) being performed in lab *as they were actually done*, not as they theoretically would be done! Also, any observations made should be noted. In general, write what you see and what you do, as you see it and as you do it. The writing should be in the passive voice as this is much more efficient and economical.

EXAMPLE:

Correct: 0.230 mL of methyl benzoate was added to a tared 5-mL reaction vial.

Incorrect: I weighed my 5-mL reaction vial. Then I went to the hood and got 0.230 mL of methyl benzoate and put it in the vial.

11. SAMPLE PAGES

Sample Abstract

Oxidation of Benzoin to Benzil

In this experiment, benzil was synthesized by the oxidation of benzoin with copper (II) acetate acting as the oxidizing agent (in the presence of ammonium nitrate). Benzoin (0.3038 g) was reacted with 2.0 mL of the oxidizing agent solution, and heated under reflux for one hour. After cooling, 0.2998 g of crude benzil product was then isolated by vacuum filtration. Recrystallization from ethanol gave 0.1631 g of pure benzil (54.2%, mp 86-88°C [lit. 95-96°C]).

24

Date: March 2, 2008
Lab Section: Tu AM

Title: Oxidation of Benzoin to Benzil

Reference: Cerny/Schwartz pp. 111–115
and www.chemfinder.com

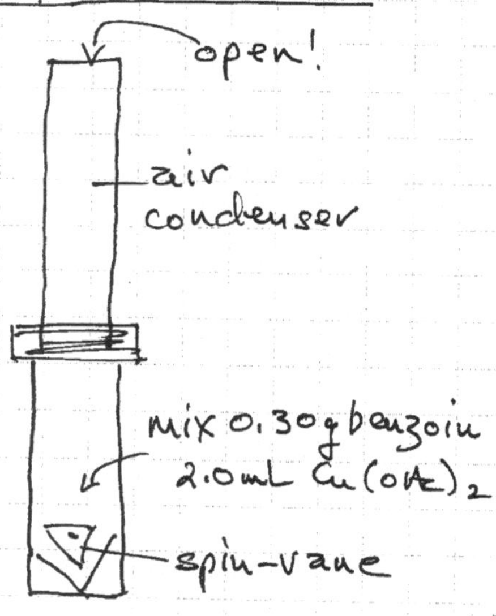

benzoin $\xrightarrow[\text{NH}_4\text{NO}_3]{\text{Cu(OAc)}_2}$ benzil

Table of Physical Constants

Reagent	MW (g/mol)	mp °C	bp °C	density (g/mL)
Benzoin	212.25	137	–	–
Benzil	210.23	95	–	–
Ethanol	46.07		78.3	0.789

Note: expected color change
$Cu^{+2} \rightarrow Cu^{+1}$
blue $\rightarrow$ green

25

Date: ___________
Lab Section: ____________

Written BEFORE your lab

Oxidation Procedure

1) Place 0.30 g. benzoin in 5 mL vial — add 2.0 mL $Cu(OAc)_2$/NH_4NO_3 soln — add spin-vane & attach air condenser
2) set-up reflux — with stirring until solid dissolves AND turns green — then reflux for 1 hr.
3) cool and remove condenser
4) add 4 mL cold H_2O to a small beaker
5) Transfer reaction mixture into the beaker. Rinse vial w/ 1 mL H_2O & add to beaker
6) stir vigorously w/ glass-rod to induce crystals & break-up solid
7) vac. filter w/ Hirsch funnel
8) wash crude benzil w/ 5 mL cold H_2O. Then vacuum dry for ~5 min.
*9) weigh crude product
10) transfer product to 10 mL Erl. flask — add HOT, BOILING EtOH (~ 7mL EtOH/ 1 g. benzil) — swirl until dissolved
11) slowly cool to rt → yellow crystals form
12) AFTER rt → ice-bath (20 min)

Written DURING your lab

3/6/2007

1) 0.3038 g benzoin + 2.0 mL $Ca(OAc)_2$ spin-vane ✓ spin ⓐ 3.3 speed after attaching air condenser
2) heated under reflux — used Al block. Solid dissolved ~ 2 min → green solution. Continued heating ~ 61 min ⟵ boiling!
3) cooled to rt — transferred to 4 mL H_2O in beaker (H_2O was ice-cold)
5) rinsed vial + spin-vane w/ 1 mL H_2O — added to beaker
6) stirred w/ glass-rod — yellowish solid formed
7) vac. filtered + washed product w/ H_2O
9) wt vial + cap + benzil   9.9282
    wt vial + cap   9.6284
      crude benzil → 0.2998 g.
10,11) recrystallized using boiling EtOH allowed to cool to RT → solid BUT not needles; so I redissolved solid allowed to cool more slowly. AH-HA! yellow needle-like solid!
12) into ice-bath for 25 min
13) vac. filtered → broken yellow needle-like solid

Date: _______________
Lab Section: _______________

13) vac. filter
14) rinse flask w/ice-cold EtOH
    & pour over crystals in funnel
15) vac. dry crystals for ~ 10 min
16) air-dry crystals until next
    lab period
17) weigh benzil
18) take mp (pure = 95°C)
    expected: 92-94°C

3/6/2007

14) rinsed w/2 mL ice-cold EtOH
15) vac. dry for 8 min. → spread
solid out on watchglass to air
dry until next lab.

3/13/2007

vial+cap+benzil   9.7043
vial+cap   9.5412
       wt benzil   0.1631 g.

% Yield:

0.3038g. benzoin → 0.3008g benzil
                      if 100%

$\dfrac{0.1631}{0.3009} \times 100 = 54.2\%$ pure

Note $\dfrac{0.2988}{0.3009} \times 100 = 99.6\%$ for the
                               crude!

mp = 86-88°C

Conclusion: I need to do a better job on
recrystallization!
mp is low! BUT the actual
solid "looks good." Re-do mp
more slowly.

A Guide to the Chemical Literature

For each experiment you are asked to go beyond the chemical information provided in the experimental write-up. For instance, you need to know the melting points for all solids and the boiling points for all liquids. To find this kind of information (i.e., physical constants: bp, mp, density, color, crystalline form, solubilities—even molecular weight) one turns to handbooks such as the *CRC Handbook of Chemistry and Physics, the Dictionary of Organic Compounds, Lange's Handbook of Chemistry, and The Merck Index.*

You also should be aware that chemical catalogs [for instance: *Aldrich Catalog: Handbook of Fine Chemicals*] are a possible source of this information for those chemicals which that concern supplies.

Always available in the laboratory are the *CRC Handbook and the Aldrich Catalog* for your use. A very helpful online chemical database is available at http://www.chemspider.com or https://www.sigmaaldrich.com/chemistry.html (use the search box in the upper right)

WHAT IS GREEN CHEMISTRY?

Green Chemistry is the utilization of a set of principles that reduces or eliminates the use or generation of hazardous substances in the design, manufacture and application of chemical products.

Green chemistry is a revolutionary philosophy that seeks to unite government, academic and industrial communities by placing more emphasis on tending to environmental impacts at the earliest stage of innovation and invention. This approach requires an open and interdisciplinary view of materials design, applying the principle that it is better to not generate waste in the first place, rather than disposing or treating it afterwards. Environmentally benign alternatives to current materials and technologies must be systematically introduced across all types of manufacturing.

Currently, environmentally benign alternative technologies have proven to be economically superior and function as well or better than more toxic traditional options. When hazardous materials are cut from processes, all hazard-related costs are cut as well, significantly reducinghazardous materials handling, transportation, disposal and compliance concerns.

Given a choice between traditional options and green solutions, business leaders choose responsibly. Unfortunately there is a significant shortage of more responsible green alternatives. Scientists and non-scientists alike can begin to address this technological gap by recognizing the interconnectivity between the construction of materials and environmental protection. There is tremendous untapped opportunity for ingenuity and reward at the chemical design stage; this is the central concern of Green Chemistry.

In the early 1990's, in response to the Clinton/Gore administration's reinventing government initiative, the US EPA's Office of Pollution Prevention and Toxics coordinated the development of the Presidential Green Chemistry Challenge Award Program. By combining elements of the EPA's design for the environment and the NSF's Technology of a sustainable environment, Green Chemistry as a philosophy was born.

While there are many programs nationally and internationally that focus on pollution prevention, sustainability and environmental issues, Green Chemistry places unique emphasis on practicing pollution prevention at the earliest stages of the design process.

The Presidential Green Chemistry Challenge Award was established as an incentive to practice, promote and disseminate Green Chemistry in a non-regulatory fashion. This remains the only Presidential award given for chemistry in the United States.

In 1998, the 12 Principles of Green Chemistry were articulated in the book "Green Chemistry: Theory and Practice". These principles reflect ongoing activities by individuals in academia, industry and government to reduce or eliminate the use and/or generation of hazardous materials and chemical processes.

12 PRINCIPLES OF GREEN CHEMISTRY*

1. *Prevention*—It is better to prevent waste than to treat or clean up waste after it has been created.
2. *Atom Economy*—Synthetic methods should be designed to maximize the incorporation of all materials used in the process into the final product.
3. *Less Hazardous Chemical Synthesis*—Wherever practicable, synthetic methods should be designed to use and generate substances that possess little or no toxicity to human health and the environment.
4. *Designing Safer Chemicals*—Chemical products should be designed to effect their desired function while minimizing their toxicity.
5. *Safer Solvents and Auxiliaries*—The use of auxiliary substances (e.g., solvents, separation agents, etc.) should be made unnecessary wherever possible and innocuous when used.
6. *Design for Energy Efficiency*—Energy requirements of chemical processes should be recognized for their environmental and economic impacts and should be minimized. If possible, synthetic methods should be conducted at ambient temperature and pressure.
7. *Use of Renewable Feedstocks*—A raw material or feedstock should be renewable rather than depleting whenever technically and economically practicable.
8. *Reduce Derivatives*—Unnecessary derivatization (use of blocking groups, protection/deprotection, temporary modification of physical/chemical processes) should be minimized or avoided if possible, because such steps require additional reagents and can generate waste.
9. *Catalysis*—Catalytic reagents (as selective as possible) are superior to stoichiometric reagents.
10. *Design for Degradation*—Chemical products should be designed so that at the end of their function they break down into innocuous degradation products and do not persist in the environment.
11. *Real-Time Analysis for Pollution Prevention*—Analytical methodologies need to be further developed to allow for real-time, in-process monitoring and control prior to the formation of hazardous substances.
12. *Inherently Safer Chemistry for Accident Prevention*—Substances and the form of a substance used in a chemical process should be chosen to minimize the potential for chemical accidents, including releases, explosions, and fires.

---

* Anastas, P. T.; Warner, J. C. Green Chemistry: Theory and Practice, Oxford University Press: New York, 1998, p. 30.

ORGANIC LAB CHECK-IN/OUT SHEET

Name _____________________   Student # _____________________

Check-In (Date) _____________________   Check-Out (Date) _____________________

Instructor's Signature (In) _____________________   Instructor's Signature (Out) _____________________

Lab Day: Mon Tues Wed Thurs Fri [AM PM]   Drawer # _____________________

EQUIPMENT LIST

Item	In	Out	Item	In	Out
Microkit (includes the following items):			Small test tubes (10 × 75) – 10		
5-mL reaction vials – 2			Large test tubes (18 × 150) – 3		
3-mL reaction vial – 1			15-mL centrifuge tubes – 1		
Teflon spin vane – 2			Assorted watch glasses – 2		
10-mL rb flask – 1			Thermometers – 1		
Air condenser – 1			Pasteur pipet bulbs – 4		
Jacketed condenser – 1			Rubber septum – 1		
Drying tube – 1			Wash bottle – 1		
Thermometer adapter – 1			Brushes (large & small)		
Vacuum adapter – 1			Test tube holder – 1		
Claisen head – 1			Test tube block – 1		
Distilling head – 1			Forceps (tongs & tweezers)		
10-mL beakers – 2			Plastic dishpan		
30-mL beakers – 2			Spatulas (large & small)		
50-mL beakers – 2			Rubber tubing – 2 pieces		
100-mL beaker – 1			Vacuum tubing – 2 pieces		
150-mL beaker – 1			Neoprene adapters – 2		
250-mL beaker – 1			3-finger clamps – 2		
10-mL Erlenmeyer flasks – 2			Glass stirring rods – 2		
25-mL Erlenmeyer flasks – 2			Capillary tubes (closed end)		
50-mL Erlenmeyer flask – 1			25-mL filter flask – 1		
10-mL graduated cylinder – 1			Hirsch funnel w/frit (micro) – 1		
25-mL graduated cylinder – 1			Powder funnel – 1		

ORGANIC LAB CHECK-IN/OUT SHEET

Name ________________________ Student # ________________________

Check-In (Date) ________________________ Check-Out (Date) ________________________

Instructor's Signature (In) ________________________ Instructor's Signature (Out) ________________________

Lab Day: Mon Tues Wed Thurs Fri [AM PM] Drawer # ________________________

EQUIPMENT LIST

Item	In	Out	Item	In	Out
Microkit (includes the following items):			Small test tubes (10 × 75) – 10		
5-mL reaction vials – 2			Large test tubes (18 × 150) – 3		
3-mL reaction vial – 1			15-mL centrifuge tubes – 1		
Teflon spin vane – 2			Assorted watch glasses – 2		
10-mL rb flask – 1			Thermometers – 1		
Air condenser – 1			Pasteur pipet bulbs – 4		
Jacketed condenser – 1			Rubber septum – 1		
Drying tube – 1			Wash bottle – 1		
Thermometer adapter – 1			Brushes (large & small)		
Vacuum adapter – 1			Test tube holder – 1		
Claisen head – 1			Test tube block – 1		
Distilling head – 1			Forceps (tongs & tweezers)		
10-mL beakers – 2			Plastic dishpan		
30-mL beakers – 2			Spatulas (large & small)		
50-mL beakers – 2			Rubber tubing – 2 pieces		
100-mL beaker – 1			Vacuum tubing – 2 pieces		
150-mL beaker – 1			Neoprene adapters – 2		
250-mL beaker – 1			3-finger clamps – 2		
10-mL Erlenmeyer flasks – 2			Glass stirring rods – 2		
25-mL Erlenmeyer flasks – 2			Capillary tubes (closed end)		
50-mL Erlenmeyer flask – 1			25-mL filter flask – 1		
10-mL graduated cylinder – 1			Hirsch funnel w/frit (micro) – 1		
25-mL graduated cylinder – 1			Powder funnel – 1		

PART II

THE TECHNIQUES

DISTILLATION

The origins of distillation are lost in antiquity, as humans, in their thirst for more potent beverages, found that dilute solutions of alcohol from fermentation could be separated into alcohol-rich and water-rich portions by heating the solution to boiling and condensing the vapors above the boiling liquid—the process of distillation. Since ethyl alcohol (ethanol) boils at 78°C and water boils at 100°C, one might naively assume that heating a 50:50 mixture of ethanol and water to 78°C would cause the ethanol molecules to leave the solution as a vapor that could be condensed to pure ethanol. Such is not the case. In a mixture of 50:50 ethanol and water, the water boils near 87°C, and the vapor above it is not 100% ethanol.

Consider a better-behaved mixture: cyclohexane and toluene. The vapor pressures as a function of temperature are plotted in Figure 1. When the vapor pressure of the liquid equals the applied pressure, the liquid boils, so this diagram shows that at 760 mm pressure—standard atmospheric pressure—these pure liquids boil at about 78°C and 111°C, respectively. If one of these pure liquids were to be distilled, we would find that the boiling point of the liquid would equal the temperature of the vapor and that the temperature of the vapor would remain constant throughout the distillation.

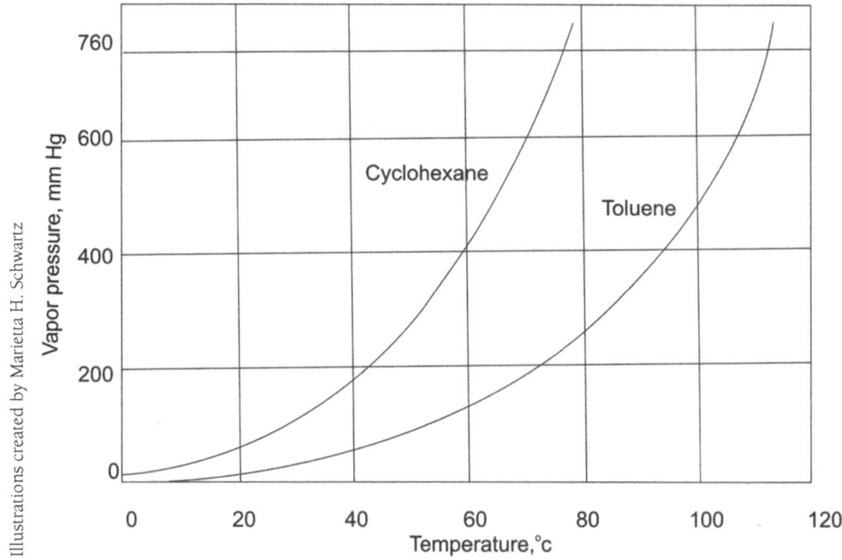

Figure 1—*Vapor pressure versus temperature for cyclohexane.*

Figure 2 is a boiling-point-composition diagram for the cyclohexane-toluene system. If a mixture of 75 mole percent toluene and 25 mole percent cyclohexane is heated, we find that it boils at 100°C, or point A (see Figure 2). Above a binary mixture of cyclohexane and toluene, the vapor pressure has contributions from each component. Raoult's law states that the vapor pressure of the cyclohexane is equal to the product of the vapor pressure of pure cyclohexane and the mole fraction of cyclohexane in the liquid mixture:

$$P_c = P°_c N_c$$

where P_c is the partial pressure of cyclohexane, $P°_c$ is the vapor pressure of cyclohexane at the given pressure, and N_c is the mole fraction of cyclohexane in the mixture. It is similar for toluene:

$$P_t = P°_t N_t$$

The total vapor pressure above the solution, P_{Tot}, is given by the sum of the partial pressures due to cyclohexane and toluene:

$$P_{Tot} = P_c + P_t$$

Dalton's law states that the mole fraction[1] of cyclohexane in the vapor at a given temperature is equal to the partial pressure of the cyclohexane at that temperature divided by the total pressure:

$$X_c = P_c/\textbf{total vapor pressure}$$

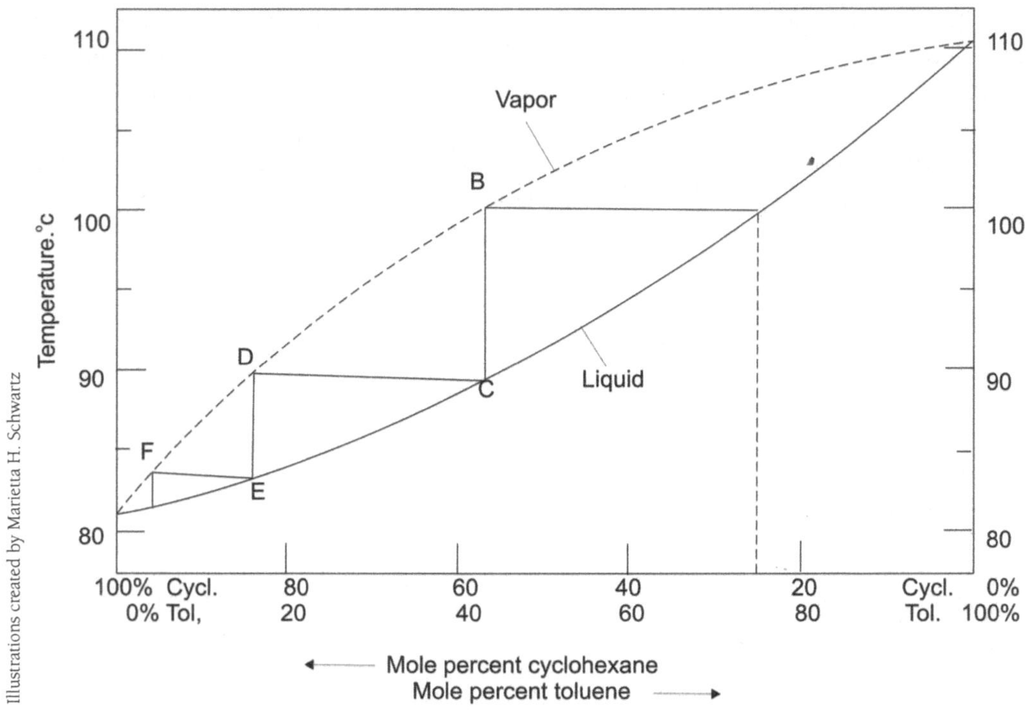

Figure 2—*Boiling point–composition curves for cyclohexane–toluene mixture.*

At 100°C, cyclohexane has a partial pressure of 433 mm and toluene a partial pressure of 327 mm; the sum of the partial pressures is 760 mm, and so the liquid boils. If some of the liquid in equilibrium with this boiling mixture was condensed and analyzed, it would be found to be 433/760 ,or 57 mole percent cyclohexane (Figure 2, point B). This is the best separation that can be achieved on a simple distillation of this mixture. As the simple distillation proceeds, the boiling point of the mixture moves toward 111°C along the line from A, and the vapor composition becomes richer in toluene as it moves from B to 110°C. To obtain pure cyclohexane, it would be necessary to condense the liquid at B and redistill it. When this is done, it is found that the liquid boils at 90°C (point C), and the vapor equilibrium with this liquid is about 85 mole percent cyclohexane (point D). Therefore, to separate a mixture of cyclohexane and toluene, a series of fractions would be collected and each of these partially redistilled. If this fractional distillation were done enough times, the two components could be separated.

This series of redistillations can be done "automatically" in a fractionating column. Perhaps the easiest to understand is the bubble-plate column used to distill crude oil fractionally. These columns dominate the skylines of oil refineries, some as much as 150 feet high and capable of distilling 200,000 barrels of crude oil per day. The crude oil enters the column as a hot vapor (see Figure 3). Some of this vapor with high-boiling components condenses on one of the plates. The more volatile substances travel through the bubble cap to the next higher plate, where some of the less volatile components condense. As high-boiling liquid material accumulates on a plate, it descends through the overflow pipe to the next lower plate, while the vapor rises through the bubble cap to the next higher plate. The temperature of the vapor rising through the cap is above the boiling point of the liquid on that plate. As bubbling takes place, heat is exchanged, and the less volatile components on that plate vaporize and go on to the next plate. The composition of the liquid on a plate is the same as that of the vapor coming from the plate below. So on each plate, a simple distillation takes place. At equilibrium, vapor containing low-boiling material is ascending, and high-boiling liquid is descending through the column.

Figure 2 shows that the condensations and redistillations on a bubble-cap column consisting of two plates correspond to moving on the boiling-point-composition diagram from point A to point E.

In the laboratory, the successive condensations and distillations that occur in the bubble-plate column take place in a distilling column. The column is packed with some material on which heat exchange between ascending vapors and descending liquid can take place. A large surface area for this packing is desirable, but the packing cannot be so dense that pressure changes take place within the column causing nonequilibrium conditions. Also, if the packing has a very large surface area, it will absorb (holdup) much of the material being distilled.[3] A number of different packings for distilling columns have been tried: glass beads, glass helices, carborundum chips, etc. One of the best packings is steel wool (often purchased as a steel sponge). It is easy to put into the column, does not come out of the column as beads do, and has low surface area, good heat-transfer characteristics, and low holdup. It can be used in both microscale and macroscale apparatus.

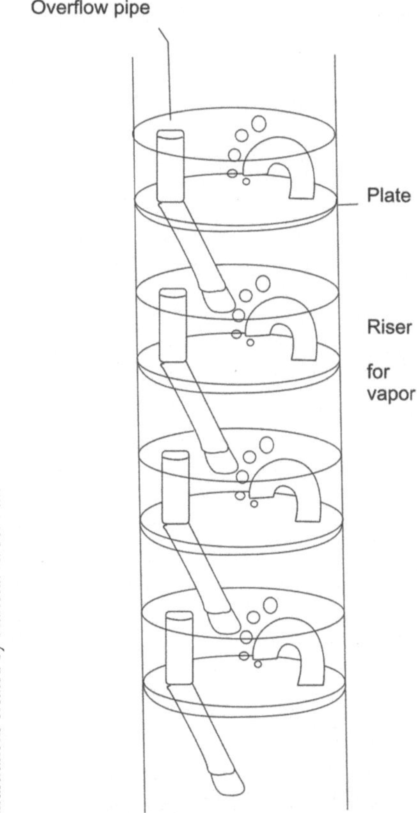

Figure 3—*Bubble plate distilling column*

The efficiency of a distillation apparatus for the separation of two materials with differing boiling points is evaluated by calculating the number of theoretical plates. Each theoretical plate corresponds to one distillation and condensation cycle, as discussed above.

Although not obvious, the most important variable contributing to a good fractional distillation is the rate at which the distillation is carried out. A series of simple distillations takes place within a fractionating column, and it is important that complete equilibrium be attained between the ascending vapors and the descending liquid. This process is not instantaneous. It should be an adiabatic process ; that is, heat should be transferred from the ascending vapor to the descending liquid with no net loss or gain of heat. In larger, more complex distilling columns, a means is provided for adjusting the ratio between the amount of material that boils up and condenses (refluxes) and is returned to the column (thus allowing equilibrium to take place), and the amount that is removed as distillate. A reflux ratio of 30:1 or 50:1 would not be uncommon for a forty-plate column; distillation would take several hours.

Carrying out a fractional distillation on the truly microscale (<1 mg) is impossible, and it is even impossible on a small scale (10–400 mg). It can be carried out on a 4 mL scale. Distillation is the best method of separating more than a few grams of liquids.

Azeotropes

Not all liquids form ideal solutions and adhere to Raoult's law. Ethanol and water are such liquids. Because of molecular interaction, a mixture of 95.5% (by weight) of ethanol and 4.5% of water boils below (78.15°C) the boiling point of pure ethanol (78.3°C). Thus, no matter how efficient the distillation apparatus, 100% ethanol cannot be obtained by distillation of a mixture of, say, 75% water and 25% ethanol. A mixture of liquids of a certain definite composition that distills at a constant temperature without change in composition is called an azeotrope; 95% ethanol is such an azeotrope. The boiling-point-composition curve for the ethanol-water mixture is seen in Figure 4. To prepare 100% ethanol, the water can be removed chemically (with calcium

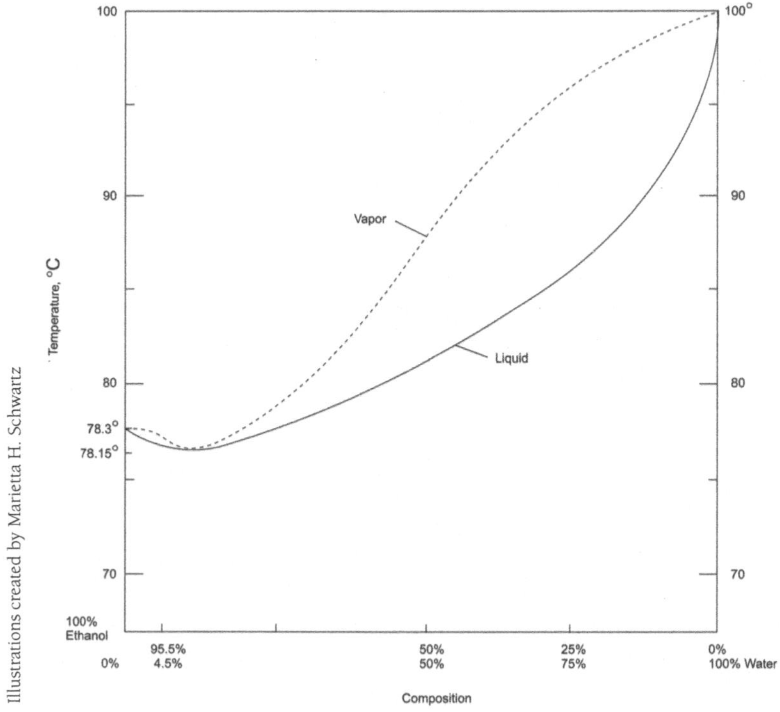

Figure 4—*Boiling point-composition curves for ethanol-water mixture.*

oxide) or as an azeotrope (with still another liquid). An azeotropic mixture of 32.4% ethanol and 67.6% benzene (bp 80.1°C) boils at 68.2°C. A ternary azeotrope (bp 64.9°C) contains 74.1% benzene, 18.5% ethanol, and 7.4% water. Absolute ethanol (100% ethanol) is made by addition of benzene to 95% ethanol and removal of the water in the volatile benzene-water-ethanol azeotrope.

Ethanol and water form a minimum-boiling azeotrope. Other substances, such as formic acid (bp 100.7°C) and water (bp 100°C), form maximum-boiling azeotropes. For these two compounds, the azeotrope boils at 107.3°C.

A pure liquid has a constant boiling point. A change in boiling point during distillation is an indication of impurity. The converse proposition, however, is not always true, and constancy of bojling point does not necessarily mean that the liquid consists of only one compound. For instance, two miscible liquids of similar chemical structure that boil at the same temperature individually will have nearly the same boiling point as a mixture. And as noted previously, azeotropes have constant boiling points that can be either above or below the boiling points of the individual components.

[1] Adapted from Williamson, K.L., "Organic Experiments," 9th Ed, Houghton-Mifflin Co., Boston, 2004, pp 65–70.

[2] The mole fraction of a component in a mixture is equal to the moles of that component, divided by the total number of moles in the mixture.

[3] Holdup: unrecoverable distillate that wets the column packing material.

DRYING AGENTS

Organic solvents used for extraction dissolve not only the compound being extracted but also water. Evaporation of the solvent then leaves the desired compound contaminated with water. Therefore, organic solvents are routinely dried (the term refers to the removal of water specifically, not to the removal of liquid) using a variety of chemical drying agents. Many choices of chemical drying agents are available for this purpose, and the choice of which one to use is governed by four factors:

- The possibility of reaction with the substance being extracted
- The speed with which it removes water from the solvent
- The efficiency of the process
- The ease of recovery from the drying agent

Some very good but specialized and reactive drying agents are potassium hydroxide (KOH), anhydrous potassium carbonate (K_2CO_3), sodium metal ($Na°$), calcium hydride (CaH_2), lithium aluminum hydride ($LiAlH_4$), and phosphorus pentoxide (PO_5). Substances that are essentially neutral and unreactive and are widely used as drying agents include anhydrous calcium sulfate ($CaSO_4$, commercially available as Drierite), magnesium sulfate ($MgSO_4$), molecular sieves, calcium chloride ($CaCl_2$), and sodium sulfate (Na_2SO_4).

Drierite, a specially prepared form of calcium sulfate, is a fast and effective drying agent. However, it is difficult to ascertain whether enough has been used. An indicating type of Drierite is impregnated with cobalt chloride, which turns from blue to red when it is saturated with water. This works well when gases are being dried, but it should not be used for liquid extractions because the cobalt chloride dissolves in many protic solvents.

Magnesium sulfate is also a fast and fairly effective drying agent, but it is so finely powdered that it always requires careful filtration for removal.

Molecular sieves are sodium alumino-silicates (zeolites) that have well-defined pore sizes. The 4A size adsorbs water to the exclusion of almost all organic substances and is a fast and effective drying agent but, like Drierite, it is impossible to ascertain by appearance whether enough has been used. Molecular sieves in the form of 1/16-in. pellets are often used to dry solvents by simply adding them to the container.

Calcium chloride, recently available in the form of pellets (4 to 80 mesh), is a very fast and effective drying agent. It has the advantage that it clumps together when excess water is present so that it is possible to know how much to add by observing its behavior. Unlike the older granular form, the pellets do not disintegrate to give a fine powder. These pellets are admirably suited to microscale experiments where the solvent is removed from the drying agent with a Pasteur pipet. It is much faster and far more effective than anhydrous sodium sulfate. But calcium chloride reacts with some alcohols, phenols, amides, and some carbonyl-containing compounds.

Sodium sulfate is another option. It has a very high capacity for water but is slow and not very efficient in the removal of water. It, like calcium chloride pellets, clumps together when wet, and solutions are easily removed from it using a Pasteur pipet. It has been used extensively in the past and should still be used for compounds that react with calcium chloride.

When using a "clumping" type of drying agent, add a small amount of the drying agent to the flask containing the wet solution. Swirl the solution and observe the clumping activity. If the entire amount of drying agent clumps together, add another small amount of drying agent. Repeat this sequence until, upon swirling, it appears to "snow" inside the flask. At this point you have added enough drying agent. Allow the flask to sit undisturbed for at least ten minutes; then remove the now-dry solution from the drying agent with a Pasteur pipet and place it in a new, dry container for evaporation.

EXTRACTION

BACKGROUND

Extraction is a very common method for the separation of an organic compound from a reaction mixture or from a solid mixture. Brewing a cup of tea is an example of a solid-liquid extraction: the flavoring compounds and caffeine are "extracted" into the hot water from the solid tea. These compounds dissolve easily in the hot water and are therefore removed from the solid mixture. In this example, it is obvious that the separation is based on the water solubility of those compounds being removed from the solid tea mixture. In liquid-liquid extraction, the desired compound (solute) will also be distributed within each layer of two solvents based on the solubility of that solute in each of the two liquids. A good solvent for the extraction of an organic compound should (1) have a high solubility for the wanted compound; (2) be immiscible with the other solvent used (to separate into two distinct layers); (3) have a low boiling point (since this liquid needs to be removed later); and (4) be nonreactive.

Water is most often one of the two solvents used. Typical solvents for extraction are listed in the table below. Solvents with a density less than water will separate as the top layer, and those with a density greater than water will separate as the lower layer.

TABLE 1—*Properties of Common Extraction Solvents*

Solvent	b.p., °C	Density, g/mL	Comments
Water	100	1.00	For polar compounds, often with acid or base
t-Butyl methyl ether	55.2	0.74	Good general solvent, absorbs some water, flammable, forms azeotrope with water (b.p. 52.6 °C)
Diethyl ether	34.5	0.71	Good general solvent, absorbs some water, very flammable, forms peroxides
Dichloromethane	40	1.34	Good general solvent, suspected carcinogen
Petroleum ether	35-60	~0.64	For nonpolar compounds, very flammable
Hexane	69	0.66	For nonpolar compounds, flammable
Toluene	111	0.87	For aromatic and nonpolar compounds, difficult to remove

Source: J. W. Lehman, *Multiscale Operational Organic Chemistry* (Upper Saddle River, NJ: Prentice-Hall, 2002), 607.

In a liquid-liquid extraction, after the two liquids are mixed, the desired compound will be found in both layers depending on its relative solubility in each.[1] To attain an efficient extraction, the desired compound should be appreciably more soluble in one of the solvents. This ratio of concentrations of the compound dissolved in each solvent is called the distribution coefficient, K. If K = 1, equal amounts of the compound will dissolve in each layer. The larger the value of K, the more effective the extraction will be. Let us consider an extraction of an organic compound Q with a diethyl ether and water system. A good approximation of the concentrations of Q in each layer uses the relative solubilities of Q in each layer. If the solubility of Q in ethyl ether is 25 g/100 mL, and the solubility of Q in water is 5.0 g/100mL, then the approximation of K is as follows:

$$K \sim \frac{\text{solubility of Q in ether}}{\text{solubility of Q in water}} = \frac{25g/100mL}{5.0g/100mL} = 5.0$$

Suppose we have a solution of 4.0 g of compound Q in 100 mL of water; if we mix this solution with 100 mL of diethyl ether, how much Q will be extracted into the ether? Let us say that x grams of Q are now dissolved in the ether; therefore (4.0 - x) is the weight of Q remaining in the water layer. Substituting in the distribution coefficient equation above gives the following:

$$K = \frac{\dfrac{x \text{ g of Q}}{100mL \text{ ether}}}{\dfrac{(4.0 - x) \text{ g of Q}}{100mL \text{ water}}} = 5.0$$

By solving for x, we find that 3.33 g of Q (83% of the original amount of Q) is now dissolved in the ether.

It is commonly written that multiple extractions with smaller volumes are more efficient than a single extraction with a larger volume of solvent. Let us test this theory by considering the original water solution, but now dividing the original 100 mL of diethyl ether into two portions of 50 mL. If we do the new calculations (changing the volumes), we can calculate that 2.9 g of Q will dissolve in the first extraction with ether, and then an additional 0.8 g of Q will dissolve in the second extraction with ether, giving a total of 3.7 g of Q (92% of the original amount of Q) removed from the water. Even this simple example supports the conclusion that multiple extractions are more efficient than a single large-volume extraction. Since most organic compounds have a distribution coefficient of this size (or larger), two or three extractions will remove almost all of the organic compound.

Most organic extraction solvents are used to extract nonpolar and moderately polar compounds from water solutions. But sometimes aqueous solutions are used to extract polar compounds from organic solutions. For example, dilute aqueous sodium hydroxide solution can be used to extract carboxylic acids from organic solvents into the aqueous layer (this occurs because the carboxylic acid reacts with the base to form the carboxylate salt—ions!—which are more soluble in the water layer than in the organic layer).

$$R - CO_2H \quad + \quad NaOH(aq) \quad ---------> \quad R - CO_2^- \, Na^+ \quad + \quad H_2O$$

And then after separating the layers, if the carboxylic acid is sufficiently insoluble in water, it can be recovered by simply acidifying the aqueous layer to precipitate the solid acid.

PROCEDURE

There are two basic techniques for extraction. Which one is used depends upon the total volume of the two immiscible layers. In microscale, for total volumes of up to about 12 mL, the extraction solutions can be contained in a 15 mL centrifuge tube and the separation of layers accomplished with a Pasteur pipet. In standard scale, for larger volumes, a separatory funnel is used. In either case, an extraction involves mixing the two phases, allowing them to separate within the original container, and then actually separating these two phases into their own containers!

Microscale: This small-volume extraction is usually done in either a 5-mL reaction vial or a 15-mL centrifuge tube, because both have a v-shaped bottom to allow for an easier separation of the two solutions. The container is capped and the liquids are shaken gently to mix the layers. The cap is carefully loosened to vent any built up pressure, and then this is repeated until the two phases are thoroughly mixed. The cap is again loosened, and the container allowed to stand undisturbed until the layers have separated.[2] The lower layer is removed from the container with a Pasteur pipet[3] and transferred to another container. Once the layers are separated, the correct layer[4] can then be extracted again as necessary. Repeat the process for an additional extraction, and then combine like layers before continuing with the procedure.

Standard Scale: This larger-volume extraction is done in a separatory funnel, which has a stopcock in the stem. This makes it possible to drain out the lower layer into a different container and retain the upper layer within the separatory funnel. It is very important that the Teflon stopcock be secured by a washer, rubber O-ring, and locknut (in that order) and that the locknut be tightened enough to prevent leaking, but still allow the stopcock to turn easily. The separatory funnel is then supported by an iron ring (clamped to a ring stand), which has been wrapped with rubber tubing as a "shock guard" (see Figure 1). Close the stopcock, and using a stemmed funnel, add the liquid to be extracted through the top. Then add the required volume of the extraction solvent into the funnel.[5] Place the glass stopper into the top opening (it must fit perfectly!). With both hands on the funnel, invert it (typically one hand cradles the stopper into the top, and the other hand holds the stopcock so that it may be opened; see Figure 2). Vent the separatory funnel by opening the stopcock to release any built-up pressure.

Warning! The stem of the funnel must be pointed up and away from you and your lab mates, because some of the contents may spray out when the pressure is released! Often a "whooshing" sound can be heard.

Close the stopcock and gently shake (or swirl) the funnel, and then again invert the funnel and open the stopcock. Continue to shake the funnel more vigorously (careful—watch for emulsions), venting as necessary, for about two minutes. Now replace the funnel on the support ring, remove the stopper, and allow the funnel to remain undisturbed until there is a sharp, dividing interface between the two layers. Open the stopcock[6] and drain out the bottom layer into an Erlenmeyer flask (see Figure 3). As the interface approaches the bottom of the funnel, partially close the stopcock to slow the drainage rate, and then quickly close the stopcock when the interface reaches it. The top layer can then be removed by simply pouring it out the top of the funnel into an appropriate container. Repeat the process for an additional extraction, and then combine like layers before continuing with the experimental procedure.

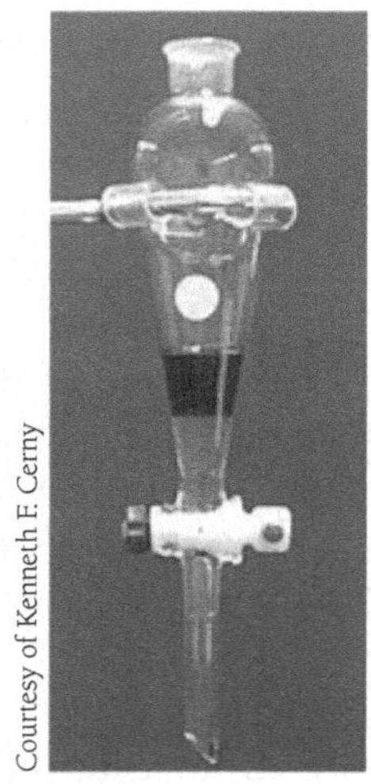

Figure 1—A separatory funnel

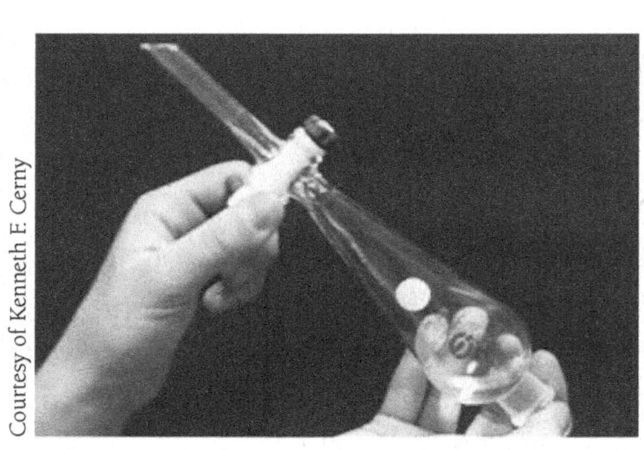

Figure 2—An inverted separatory funnel

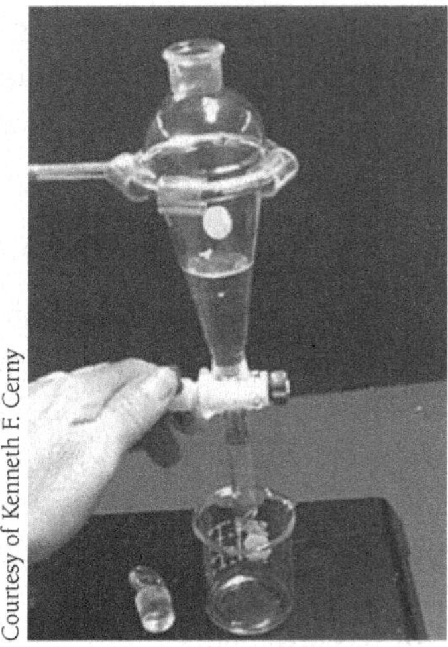

Figure 3—Draining the separatory funnel

[1] The terms "extraction" and "washing" imply slightly different operations, but the mechanics are identical. Extraction is the process of removing the wanted compound, while washing is the process of removing the impurities from the wanted compound.

[2] Sometimes the layers do not separate cleanly, either because an emulsion forms or because some liquid droplets of one phase remain in the other. An emulsion (a dispersion of very fine droplets of one immiscible liquid suspended in another) can often be broken up by (a) using a glass stirring rod to stir the liquids gently at the interface and to push droplets away from the sides or bottom of the container; or (b) by adding solid NaCl to try to make a salt saturated aqueous layer—the water molecules solvate ions very well and are then less available to solvate organic molecules. If one knows there might be a tendency to form an emulsion, the "thorough mixing" called for is done much more gently!

[3] Squeeze the rubber bulb of the pipet to remove the air. Then place the tip of the pipet at the bottom of the v-shaped container; very slowly release the bulb, and draw the bottom layer into the pipet, without also drawing up any of the top layer. Drawing the bottom layer allows the interface between the layers to come down to the narrowest part of the container, allowing for a sharp separation between the layers. It will take some practice and a steady hand to learn this technique, but it is important to do so. An accurate separation is especially important for the last separation in multiple extractions. Product loss due to spurting out of the tip of the pipet can often be eliminated by first rinsing the pipet with the solvent involved in the lower layer (to fill it with solvent vapor) and also being sure that all solutions are at room temperature or below.

[4] A rule: NEVER throw any layer away until you actually have isolated the desired compound! It is really difficult to isolate your compound from either the sink trap or the organic-waste bottle! The "correct layer" can often be determined by careful observation when adding the two liquids together. One can usually tell whether the liquid being added sinks below (or stays on top) of the existing liquid. And once the layers have been separated, one can add a small volume of water to the layer that is thought to be aqueous and confirm that indeed there is only one layer after the addition. The relative densities of the solvents is also very helpful.

[5] The total volume of both liquids should not be larger than three-quarters of the capacity of the separatory funnel. If the volume limit is exceeded, you may need to carry out the extraction in several steps and then combine layers at the end.

[6] If the liquid does not drain smoothly from the separatory funnel, check to make sure the glass stopper has been removed.

MELTING-POINT

The term "melting point" is really a misnomer, as most organic compounds actually exhibit a melting-point range. The temperature range between the first indication of melting to that where melting is complete is usually 2°C or less for a pure compound. Impure substances show a lower and wider melting-point range. One way to establish the purity of a solid is to recrystallize it until the melting point is constant; each recrystallization will raise and sharpen (make narrower) the melting-point range.

We will use a Mel-Temp apparatus to take melting points (see Figure 1). The solid sample, which must be dry, is crushed to a fine powder; the powder is then introduced into the open end of a melting-point capillary tube by gently pressing the open end into a small pile of the powder (see Figure 2).

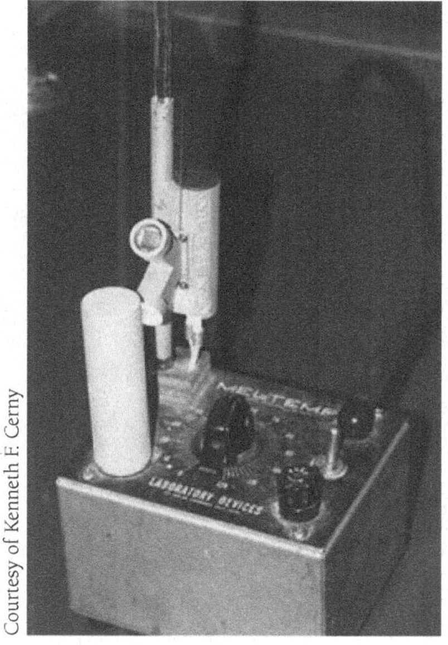

Figure 1—*The Mel-Temp apparatus*

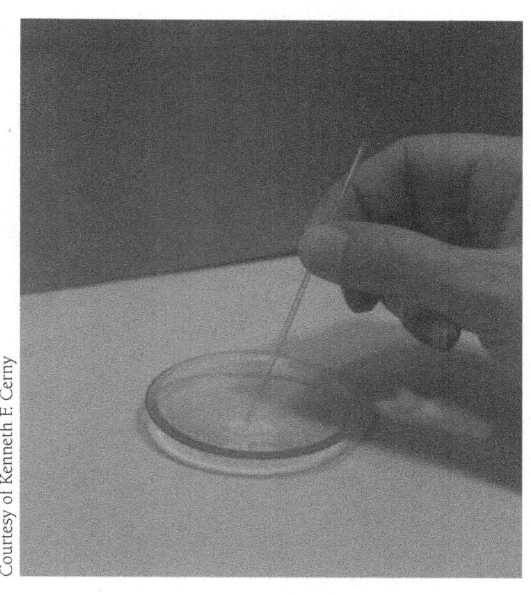

Figure 2—*Preparing a sample for melting-point determination*

Upon inverting the capillary and gently tapping it vertically on the bench top, the solid will fall to the closed end.[1] Typically a depth of 1 to 2 mm of solid is all that is necessary (since there is a magnifier in the melting-point apparatus). The capillary is placed into one of the slots[2] in the Mel-Temp, and heat is applied by varying the voltage control.[3] The rate of heating must be slow—near the melting point the temperature should rise about 2°C per minute! When the first sign of melting occurs, note that temperature; and when the entire sample has melted, again note that temperature.[4] This range of temperatures is the melting point of the solid, which you should record in your laboratory notebook.[5]

[1] Some solids need more vigorous "tapping" to move to the closed end of the capillary. Dropping them down a four-foot length of glass tubing held vertically often does the trick.

[2] Careful: the unit could be "hot" from a previous melting point! Look at the thermometer to be sure that the temperature is at least 20°C below the anticipated melting point for your solid. You may need to wait until the apparatus cools or use a different one.

[3] You may heat quickly (with the voltage control between 40 and 50) until the temperature gets to about 20°C below the expected melting point of your solid. Then turn the voltage control to 0 (off), and wait for the thermometer to equilibrate. Now continue to heat so the temperature rises very slowly until the solid melts.

[4] Noting the temperature change will require you to look away from the solid to the thermometer scale and then return to the eyepiece to again view your solid sample. This takes a finite amount of time, so there should be little, if any, increase in the temperature during that time.

[5] You may throw away the capillary tube (in the glass-disposal container!). Remember to turn the Mel-Temp **OFF** and to turn the voltage control to 0.

RECRYSTALLIZATION

BACKGROUND

The easiest and most useful method to purify solids is recrystallization. We use the term **re**crystallization, because one begins with a solid that usually had already crystallized from a reaction mixture (or maybe from a separation solution). In its most simple form, recrystallization involves dissolving this original solid in a hot (usually boiling) solvent, and then allowing the resulting solution to cool (first to room temperature and then in an ice-water bath) to let the solid crystallize again from solution. Recrystallization relies on the fact that the solubility of a solid in the "recrystallization solvent" increases with temperature. So if you can get a saturated solution at the elevated temperature, the solid must come out of solution at the lower temperature. These new crystals are usually much purer than the crude solid, because most of the impurities remain dissolved in the solvent. The choice of which solvent to use is often the most difficult part of this purification technique; however, in these experiments the solvent of choice is given. In actual practice, there may be other steps (decolorization and/or gravity filtration to remove insoluble impurities) required to obtain the pure solid.

PROCEDURE

Using a 125-mL Erlenmeyer flask and hot plate, bring about 50 mL of the solvent to a gentle boil. This will be your reservoir of "hot" solvent.

Place the crude solid to be recrystallized into a small Erlenmeyer flask. The solid needs to be dissolved to make a saturated solution, so you need to add hot solvent in small amounts (about 1 mL using a Pasteur pipet) until the solid just dissolves <u>at the boiling point!</u> This will usually mean trial and error; that is, add a small amount of solvent, and heat the solution to boiling to see if the solid dissolves.[1] If it does not all dissolve, add more solvent and boil; repeat until all of the solid eventually dissolves at the boiling point of the solvent.[2] You now have a saturated solution. (Of course, this can be difficult, since low-boiling solvents will continue to "boil away"—but perseverance!)

Colored impurities can often be removed with pelletized charcoal, BUT first note that one can indeed obtain white crystals from a slightly colored recrystallization solvent! Check with your instructor.[3]

Place your Erlenmeyer flask (with the saturated solution of solid in the recrystallization solvent) onto the desktop where it can sit unattended until it reaches room temperature. Place a folded paper towel between the flask and the desktop for "insulation." Slow cooling allows the crystals to form more slowly and therefore with a higher degree of purity.

After the flask has reached room temperature, carefully place it into an ice-water bath to continue cooling. Be sure the flask does not tip over into the water! Again allow the flask to sit unattended for at least twenty minutes (to allow the solution to reach the temperature of the ice-water bath).[4]

Collect the recrystallized solid by vacuum filtration using a Hirsch funnel.

Recrystallization from Mixed Solvents

Often solids are recrystallized from a solvent pair. The solvents must be miscible in each other, and the desired compound must be very soluble in one solvent and relatively insoluble in the other solvent. In a mixed-solvent recrystallization, the solid is dissolved in the solvent in which it is most soluble, and then the other solvent is added slowly to bring the solution to the saturation point.

Most of the steps are the same as above. Place the crude solid into an Erlenmeyer flask, and add the boiling first solvent in portions, stirring until the solid dissolves. (Insoluble impurities may be present. If necessary, decolorization with pelletized charcoal can occur at this point.) Now add the other solvent in small portions, keeping the whole recrystallization solution boiling. When a persistent cloudiness appears, add the first solvent dropwise until the cloudiness disappears. Set the saturated solution aside as before to cool to room temperature, and then cool in an ice-water bath. Collect the product by vacuum filtration.

[1] Your first addition of boiling solvent should <u>not</u> dissolve all of the crude solid! You are trying to make a saturated solution—if you add too much solvent, you will already have more solvent than necessary! Do not fret if this occurs, however; simply boil off some of the solvent (until you see crystals or cloudiness), and then proceed as above.

[2] You need to be careful, since some *impurities* may not be soluble in the solvent. Do not mistake such impurities for your desired solid. Excess solvent added in trying to dissolve them will decrease your yield of purified crystals. When you get to the point that most of the solid has dissolved, you must carefully note whether the addition of another portion of the solvent continued to dissolve more of the remaining solid! If not, this solid is probably an impurity and needs to be removed by gravity filtration, using coarse, fluted filter paper and a preheated funnel. If filtration is necessary, you will need to add at least a 10% excess of solvent.

[3] If decolorization is necessary, first allow the boiling recrystallization solution to cool for several minutes, and then add a <u>small</u> amount of the decolorizing carbon. Stir the solution for a minute or two, keeping the solution just below its boiling point. You will need to add at least a 10% excess of solvent; then again bring the solution to boiling, and remove the charcoal by gravity filtration, using coarse, fluted filter paper and a preheated funnel.

[4] If no crystals form after the hot recrystallization solution is cooled, crystallization often can be induced by rubbing the end of a glass stirring rod against the <u>inside</u> wall of the flask, using an up-and-down motion with the rod tip just going into (and out of) the liquid for a few minutes. (This is sometimes called "scratching" the solution.) If crystals still do not form, you probably have used too much solvent (and therefore did not have a saturated solution). Concentrate the solution by evaporating some of the solvent until the solution just becomes cloudy; then continue boiling, add solvent until the cloudiness disappears, and repeat the cooling steps.

REFLUX

Reflux, which means "a flowing back," is a technique that has two purposes:

1. A reaction mixture can be maintained at the boiling point of the combined reagents dissolved in some nonreactive solvent; and
2. The solvent is not lost due to boiling (it falls back into the reaction vessel).

Concerning the first of these purposes, different solvents (having different boiling points) would allow the temperature of the reaction mixture to be varied. For the experiments in this text, the solvent of choice will be given.

Concerning the second of these purposes, a water-jacketed condenser is placed in the upright position on the reaction container (flask or vial; see Figure 1). Then, as the vapors rise from the boiling reaction mixture, they will be cooled in the condenser and return to the original container. To allow for maximum cooling, the cold water must enter the condenser at the lower port (closer to the reaction vessel) and exit from the top to the sink.[1] Heating too strongly may allow the boiling rate to exceed the capacity of the condenser (and then solvent vapors are lost through the top of the condenser). The temperature of a refluxing reaction mixture will be the same whether boiling vigorously or slowly. One needs to control the rate of heating to keep the position of the condensing liquid in the lower third of the condenser. Once this reaction setup has become stable (heating rate and condensing efficiency), it will not require your attention.

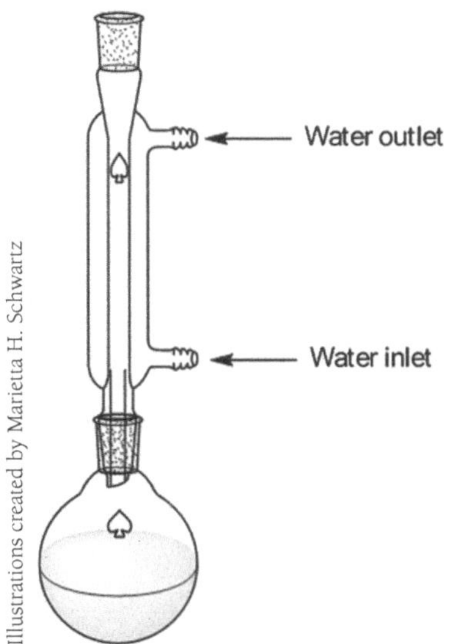

Figure 1—*Reflux setup*

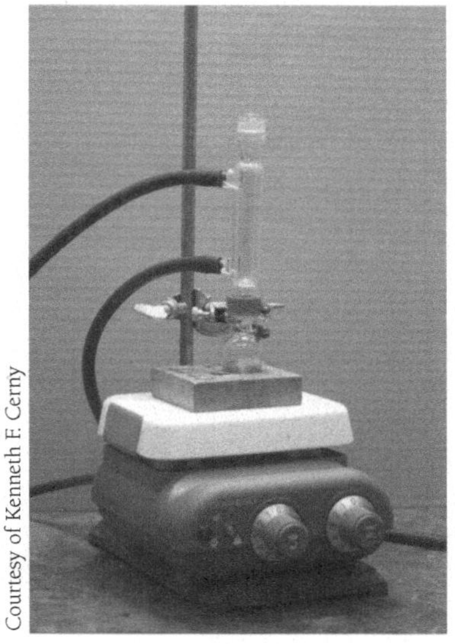

Figure 2—*Photo of reflux setup*

[1] The water hoses must be fully onto the ports! Once the water is turned on there will be pressure applied to them. You do not want to have the water squirting everywhere when a hose "pops" off!

REFRACTIVE INDICES

The refractive index n is a physical constant that, like the boiling point, can be used to characterize liquids. It is the ratio of the velocity of light traveling in air to the velocity of light moving in the liquid. It is also equal to the ratio of the sine of the angle of incidence ϕ to the sine of the angle of refraction ϕ':

n = Velocity in air/Velocity in liquid = $\sin\phi/\sin\phi'$

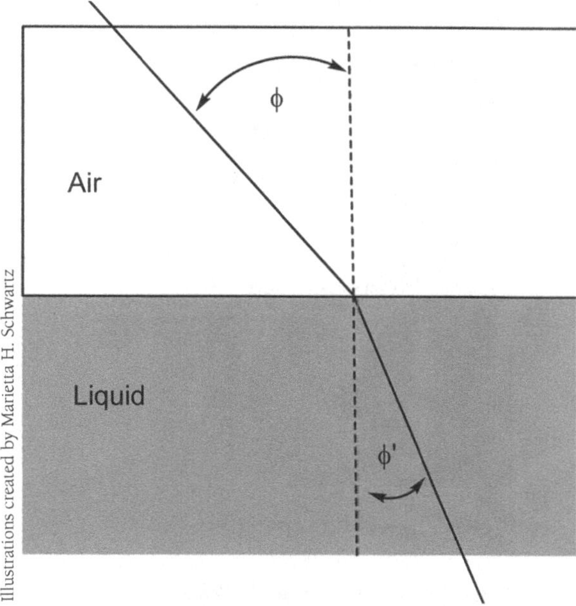

The angle of refraction is also a function of temperature and the wavelength of light (consider the dispersion of white light by a prism). Because the velocity of light in air (strictly speaking, a vacuum) is always greater than that of light through a liquid, the refractive index is a number greater than 1: for example, hexane n(20/D) = 1.3751.

The superscript "20" indicates that the refractive index was measured at 20°C, and the subscript "D" refers to the yellow D-line from a sodium vapor lamp, light with a wavelength of 589 nm.

The measurement is made on a refractometer using a few drops of liquid. Compensation is made within the instrument for the fact that white light and not sodium vapor light is used, but a temperature correction should be applied to the observed reading by adding 0.00045 for each degree above 20°C. The refractive index can be determined to one part in 10,000, but because the value is quite sensitive to impurities, across the board agreement in the literature regarding the last figure does not always exist. For this reason, most refractive indices are rounded to the nearest part per thousand.

USING THE REFRACTOMETER

The most commonly used instrument for the measurement of refractive indices is the Abbe refractometer (see Figure 1). A few drops of the sample to be analyzed are placed on the open prism. The prism is closed, and the light is turned on and positioned for maximum brightness, as seen through the eyepiece. If the refractometer is set to a nearly correct value, then a partially gray image will be seen, as shown in Figure 2a. Turn the knob so that the line separating the dark and light areas is at the crosshairs, as shown in Figure 2b. Sometimes the line separating the dark and light areas is fuzzy and colored. Turn the chromatic adjustment until the demarcation line is sharp and colorless. Then read the refractive index (Figure 2c). On a newer instrument, press a button to light up the scale in the field of vision. On older models, read the refractive index through a separate eyepiece. Read the temperature on the thermometer attached to the refractometer, and make the appropriate temperature correction to the observed index of refraction. When the measurement is complete, open the prism, and wipe off the sample with a Kimwipe, using ethanol or acetone as necessary.

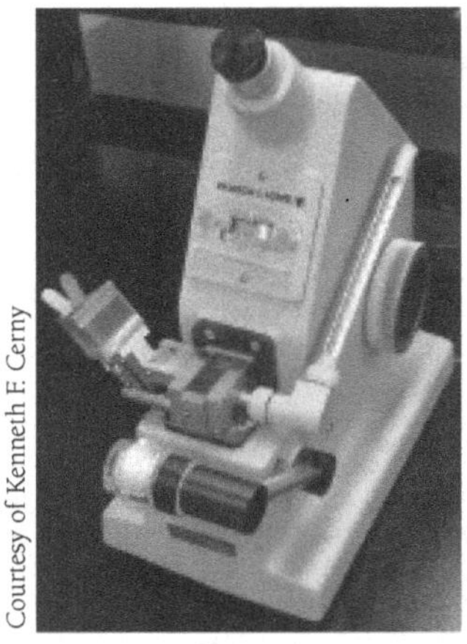

Figure 1—*Abbe refractometer*

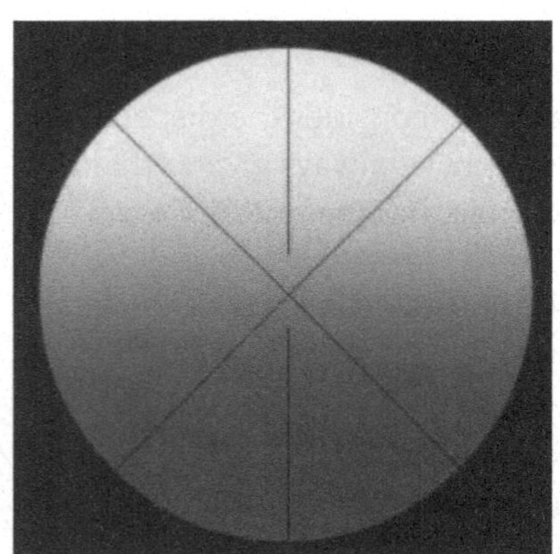

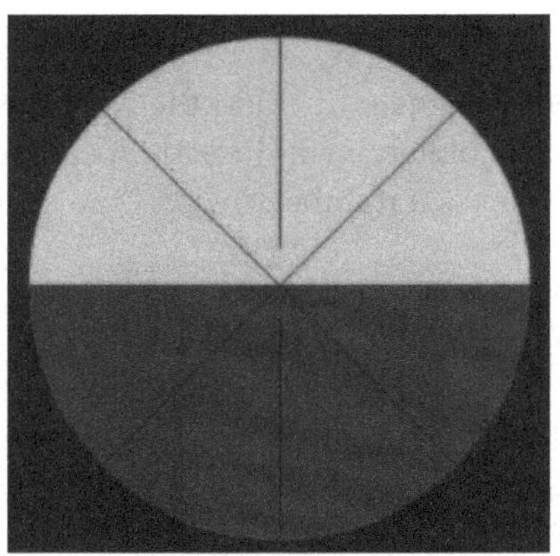

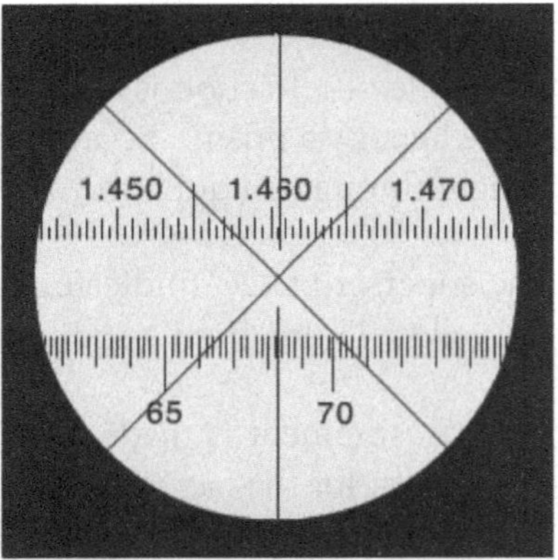

2a	2b	2c

Figure 2—*Measuring refractive indices (A), set refractometer to nearly correct value (B), focus horizon line as sharply as possible and center on the (C) read refractive index (upper scale) to four decimal places. (Pictures from John Hanson, Ph.D., Department of Chemistry, Univ. of Puget Sound).*
Source: <ins>http://www2.ups.edu/faculty/hanson/labtechniques/refractometry/smallmovie.htm</ins>
(accessed December 19, 2007)

Sublimation

Sublimation is the process whereby a solid evaporates from a warm surface and condenses on a cold surface, again as a solid. This technique is particularly useful for the small-scale purification of solids, because there is so little loss of material in transfer. If the substance possesses the correct properties, sublimation is preferred over crystallization when the amount of material to be purified weighs less than 100 mg.

Sublimation can occur readily at atmospheric pressure. For substances with lower vapor pressures, vacuum sublimation is used. At very low pressure, the sublimation becomes very similar to molecular distillation, where the molecule leaves the warm solid and passes unobstructed to a cold condensing surface and condenses in the form of a solid.

Since sublimation occurs from the surface of the warm solid, impurities can accumulate and slow down or even stop the sublimation, in which case it is necessary to open the apparatus, grind the impure solid to a fine powder, and restart the sublimation.

Sublimation is much easier to carry out on a small scale than on a large one. It would be unusual for a research chemist to sublime more than about 10 g of material, because the sublimate tends to fall back to the bottom of the apparatus.

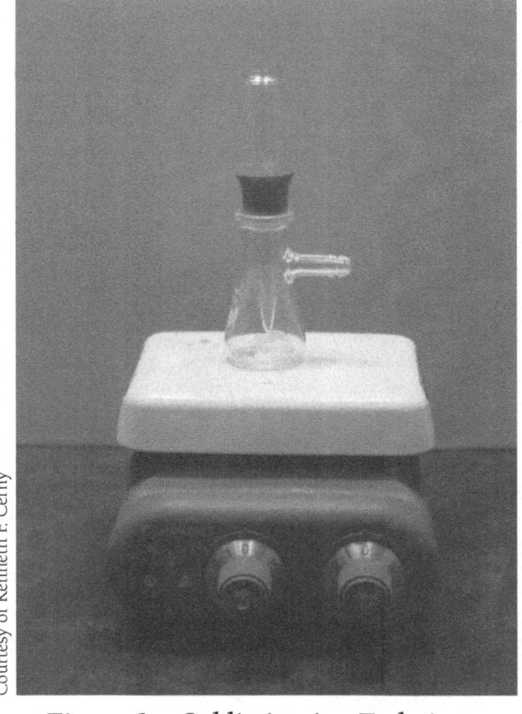

Figure 1—*Sublimination Technique*

The simplest sublimation apparatus consists simply of a centrifuge tube with a Neoprene adapter seated in the top of a filter flask, as shown in Figure 1. The impure solid is placed in the bottom of the filter flask, and the centrifuge tube is filled with ice to create a simple "cold finger." The entire apparatus is placed on a hot plate and gently heated to initiate the sublimation process. If the buildup of solid on the cold finger becomes too great, the apparatus is removed from the hot plate and disassembled. The sublimated solid is scraped onto a watch glass; the apparatus is reassembled and returned to the hot plate to continue the sublimation process.

THIN-LAYER CHROMATOGRAPHY

BACKGROUND

Thin-layer chromatography (TLC) is a sensitive, fast, simple, and inexpensive analytical technique that is commonly used during organic experiments. It is a microtechnique; as little as 10^{-9} g of material can be detected, although the usual sample size is from 1 to 100 x 10^{-6} g.

TLC involves spotting the sample to be analyzed near one end of a sheet of glass or plastic that is coated with a thin layer of an adsorbent. The sheet, which can be the size of a microscope slide, is placed on end in a covered jar containing a shallow layer of solvent. As the solvent rises by capillary action up through the adsorbent, differential partitioning occurs between the components of the mixture dissolved in the solvent and the stationary adsorbent phase. The more strongly a given component of the mixture is adsorbed onto the stationary phase, the less time it will spend in the mobile phase and the more slowly it will migrate up the TLC plate.

Uses of Thin-Layer Chromatography

1. **To determine the number of components in a mixture:** TLC affords a quick and easy method for analyzing such things as a crude reaction mixture, an extract from some plant substance, or a painkiller. Knowing the number and relative amounts of the components aids in planning further analytical and separation steps.
2. **To determine the identity of two substances:** If two substances spotted on the same TLC plate give spots in identical locations, they may be identical. If the spot positions are not the same, the substances cannot be the same. It is possible for two closely related compounds that are not identical to have the same positions on a TLC plate.
3. **To monitor the progress of a reaction:** By sampling a reaction from time to time, it is possible to watch the reactants disappear and the products appear using TLC. Thus, the optimum time to halt the reaction can be determined, and the effect of changing such variables as temperature, concentrations, and solvents can be followed without having to isolate the product.
4. **To determine the effectiveness of a purification:** The effectiveness of distillation, crystallization, extraction, and other separation and purification methods can be monitored using TLC, with the caveat that a single spot does not guarantee a single substance.
5. **To determine the appropriate conditions for a column chromatographic separation:** Thin-layer chromatography is generally unsatisfactory for purifying and isolating macroscopic quantities of material; however, the adsorbents most commonly used for TLC—silica gel and alumina—are used for column chromatography as well. Column chromatography is used to separate and purify up to a gram of a solid mixture. The correct adsorbent and solvent used to carry out the chromatography can be determined rapidly by TLC.

6. **To monitor column chromatography:** As column chromatography is carried out, the solvent is collected in a number of small flasks. Unless the desired compound is colored, the various fractions must be analyzed in some way to determine which ones have the desired components of the mixture. TLC is a fast and effective method for doing this.

Retention Factor (R_f)

The most useful measurement that can be made from a developed TLC plate is the ratio between the distances moved by the compound spot and the eluting solvent. This ratio is called the retention factor of a particular compound and is referred to as the R_f value. It is a unitless number.

Compounds that travel close to the solvent front will have an R_f value of almost 1, whereas those that do not travel very far will have R_f values that are nearer zero. R_f values by definition must fall between 0 and 1 and are reported to two decimal places. The best separations are achieved when the R_f value falls between 0.3 and 0.7. In theory, a given compound should always give the same R_f value under the given chromatographic conditions (adsorbent, eluant, temperature). It is virtually impossible to standardize these conditions, however; so Rf values for a given compound under given conditions can vary by up to 10%. R_f values are useful to anyone following a procedure, particularly if developed in conjunction with a standard material, as they indicate the likely region in which to look for the compounds on the TLC plate. They are also useful when attempting to identify unknown compounds given a range of possible knowns. For instance, the TLC plate shown below has two sets of spots: one is a known reference, while the other is a mixture of two compounds. It is reasonable to conclude that the mixture contains the known compound and one other compound.

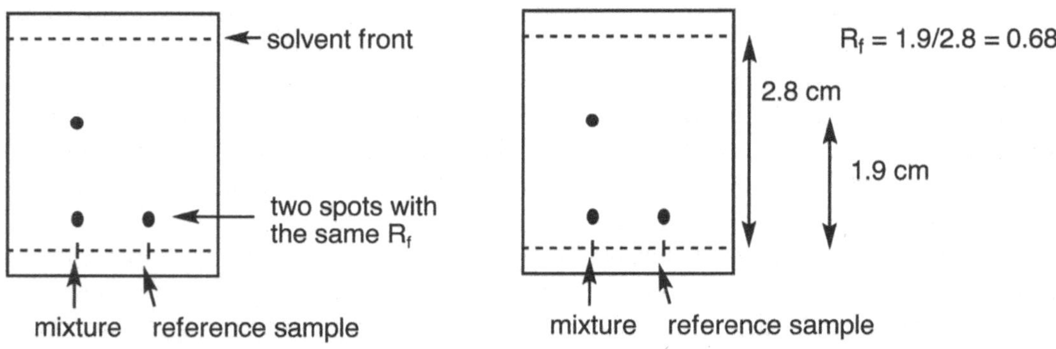

Adsorbents and Solvents

The two most common coatings for TLC plates are alumina (Al_2O_3) and silica gel (SiO_2). These are the same adsorbents most commonly used in column chromatography for the purification of macroscopic quantities of material. Of the two, alumina, when anhydrous, is the more active; that is, it will adsorb substances more strongly. It is thus the adsorbent of choice when the separation involves relatively nonpolar substrates such as hydrocarbons, alkyl halides, ethers, aldehydes, and ketones. To separate more polar substrates such as alcohols, carboxylic acids, and amines, the less active adsorbent, silica gel, is used. In an extreme situation, very polar substances on alumina do not migrate very far from the starting point (give low R_f values), and nonpolar compounds travel with the solvent front (give high R_f values) if chromatographed on silica gel. These extremes of behavior are markedly affected, however, by the solvents used to carry out the chromatography. A polar solvent will carry along with it polar substrates, and nonpolar solvents will do the same with nonpolar compounds—another example of the generalization "like dissolves like."

Table 1 lists common solvents used in chromatography, both thin-layer and column. Only the environmentally safe solvents are listed; the polarities of such solvents as benzene, carbon tetrachloride, or chloroform can be matched by other, less toxic solvents. In general, these solvents are characterized by having low boiling points and low viscosities that allow them to migrate rapidly. They are listed in order of increasing polarity. A solvent more polar than methanol is seldom needed. Often just two solvents are used in varying proportions; the polarity of the mixture is a weighted average of the two. Ligroin-ether mixtures are often employed in this way.

The order in which solutes migrate on TLC is the same as the order of solvent polarity. The largest R_f values are shown by the least polar solutes. In Table 2, the solutes are arranged in order of increasing polarity.

Table 1—*Chromatography solvents*

Solvent	bp (°C)
Petroleum ether (pentanes)	35–36
Ligroin (hexanes)	60–80
Dichloromethane	40
t-Butyl methyl ether	55
Ethyl acetate	77
Acetone	56
2-Propanol	82
Ethanol	78
Methanol	65
Water	100
Acetic acid	118

Table 2—*Order of solute migration on chromatography*

Solute	Solute
Fastest	*Slowest*
Alkanes	Ketones
Alkyl halides	Aldehydes
Alkenes	Amines
Dienes	Alcohols
Aromatic hydrocarbons	Phenols
Aromatic halides	Carboxylic acids
Ethers	Sulfonic acids
Esters	

Procedure

First prepare the TLC plate by drawing a baseline approximately 1 cm from the bottom of the plate (always use a pencil when marking a TLC plate!). Then dissolve a small quantity of sample in a few drops of a suitable solvent. Dip a small microcapillary spotter into the solution. This will cause some of the solution to rise into the tube by capillary action. Carefully touch the loaded pipet lightly onto the silica at a point marked on the baseline. This will cause some of the liquid in the pipet to be drawn onto the adsorbent, forming a small spot of your material and the solvent. The amount of material that you spot on the plate is very important; it will take practice to apply the correct amount. Too much applied material will cause the plate to be overloaded, and your spots will streak and be large. Too little material will cause the spots to be difficult to visualize after elution. The most accurate results are obtained when there is just enough sample on the TLC plate to visualize the spots after development. If you are not sure if there is enough sample spotted onto the TLC plate, put it under a UV lamp. You should be able to see the spot on the baseline as a round, dark shape.

Place the TLC plate carefully into the developing chamber. It is important to ensure that the sides of the plates are not touching the walls of the chamber and only the back is touching. The baseline where the sample is spotted must be at a level above the solvent. The solvent front will rise in a horizontal, straight line up the plate by capillary action. When the solvent front reaches a level of about 1 cm from the top of the plate, carefully remove the plate, mark the solvent front with a pencil, and allow the solvent to evaporate.

Visualization of the Chromatogram

If the substances being chromatographed are colored, then it is possible to detect the compounds visually. Colorless substances can be detected when the solvent is allowed to evaporate and the plate allowed to stand in a stoppered bottle containing a few crystals of iodine. Iodine vapor is adsorbed by the organic compound to form a brown spot. A spot should be outlined at once with a pencil, because it will soon disappear as the iodine sublimes away; a brief return to the iodine chamber will regenerate the spot.

The most useful visualization technique for TLC is observation of the plate under UV light. Analysis is quick, but one must take appropriate precautions when working with UV light. Never look directly at the UV light source, as it can be damaging to the eyes. Also, the wavelengths used for visualizing TLC plates can be damaging to the skin after prolonged exposure.

The UV light causes most aromatic molecules and molecules possessing extended conjugation to give out a bright purple fluorescence against a dark background. However, not all molecules possess this type of chromophore. Therefore, TLC plates coated with silica mixed with zinc sulfide are common in most organic labs. The zinc sulfide fluoresces green under the UV light, except where a compound on the plate quenches this fluorescence, giving a dark spot. After elution, this type of TLC plate will show dark spots on a light green background under the UV light. Each spot should be carefully outlined with a pencil. Rf values are commonly measured from the middle of the spot.

VACUUM FILTRATION

BACKGROUND

Vacuum filtration (or suction filtration) is a convenient way to separate a solid from a solid-liquid mixture. Figure 1 shows a typical setup for these experiments, using a Hirsch funnel, filter flask, and heavy-walled pressure tubing. If a larger volume of solid is being collected, the same procedure can be followed using the larger Büchner funnel (see Figure 2). Note that the Hirsch funnels have a built-in, reusable "frit" that takes the place of filter paper. Büchner funnels require a separate piece of filter paper.

Figure 1—Hirsch funnel setup

Illustrations created by Marietta H. Schwartz

Figure 2—Büchner funnel setup

PROCEDURE

Clamp the filter flask securely to a ring stand, and connect the pressure tubing to the vacuum line. Insert the Hirsch funnel into the filter flask, using a neoprene adaptor to help make a tight seal. Open the vacuum line by turning the handle to a position directly <u>above</u> the vacuum nozzle.

Add the solid-liquid slurry in portions into the Hirsch funnel while the vacuum is on. Swirl the slurry before pouring it into the funnel to facilitate transferring the solid into the funnel. Leave the vacuum on until liquid stops dripping from the funnel stem.

Often the procedure will call for washing the collected solid. To wash, first be sure that the vacuum is off, and then add the previously cooled wash liquid directly into the solid within the funnel. Carefully and gently stir the solid to suspend it in the wash liquid. Then reapply the vacuum (as before) to pull the washing liquid through the filter into the filter flask. Again leave the vacuum on until liquid stops dripping from the funnel stem, and then about three to five minutes longer. This will help to "dry" the solid.

Before turning the vacuum off, "break" the vacuum inside the apparatus by disconnecting the tubing from the side arm of the flask or by separating the funnel from the neoprene adaptor. Once the vacuum is turned off, remove the funnel from the filter flask.

Now run the tip of a spatula around the solid filter cake within the funnel. Invert the funnel over a square of glazed weighing paper (not filter paper), and carefully but solidly rap the funnel onto the paper. The solid should fall out onto the paper. Use a spatula to transfer any remaining solid from the funnel to the paper. This collected solid usually needs to air-dry until your next laboratory period; so spread it out on the paper, and store it in your desk so it will not spill.

Dispose of the filtrate ("mother liquor") as specified—usually in the organic-waste bottle. Wash the filter funnel and flask thoroughly (use a bit of acetone to get the frit clean, if necessary; you do not need to replace the frit), and return them to your laboratory drawer.

PART **III**

THE EXPERIMENTS

AMIDE SYNTHESIS: ACETANILIDE—PRELAB

Prelab Report: Due at the Beginning of the Lab Period

Name ____

Lab Section (Circle One): Mon   Tues   Wed   Thur   Fri   AM/PM

1. What kind of an amide are you making?

2. What is the nucleophile in this reaction?

3. What is the electrophile in this reaction?

4. What is the purpose of the sodium acetate?

5. What is the structure of the anilinium ion?

6. Provide the requested physical properties:

Boiling point, aniline:

Density, aniline:

Melting point, acetanilide:

Molecular weight, sodium acetate trihydrate:

AMIDE SYNTHESIS: ACETANILIDE

REACTION

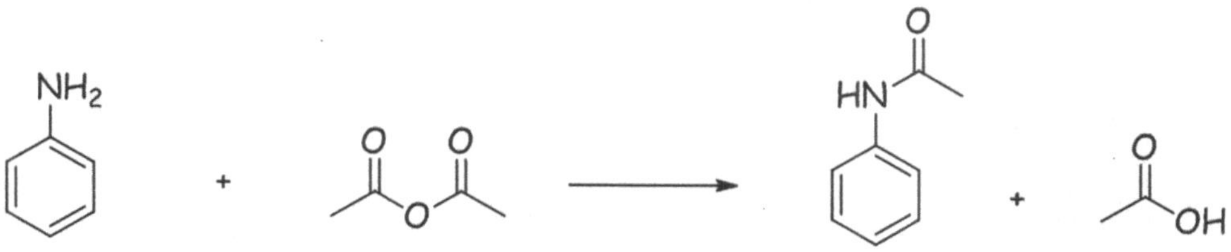

DISCUSSION

A number of important chemical and biochemical sequences are begun by the addition of nucleophilic nitrogen to an electrophilic carbonyl carbon, forming an amide functionality. Amides are classified as primary, secondary, or tertiary, depending on the number of carbon atoms attached to the nitrogen. Amides are found in such diverse compounds as penicillin V (a secondary amide) and polypeptides (amino acid sequences linked by amide bonds). Nylon is another important polymeric amide.

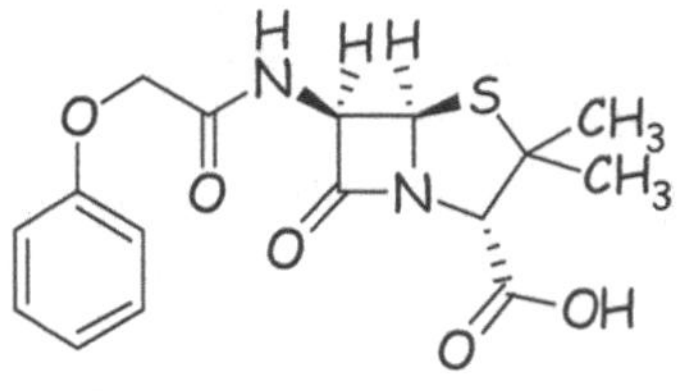

Penicillin V

This experiment illustrates the preparation of a secondary amide. The process involves the attack of a primary amine on the acetyl group of acetic anhydride. (Ammonia and secondary amines react readily with acetic anhydride as well, giving primary and tertiary amides, respectively.) The mechanism shown here is a classic example of acyl substitution (nucleophilic attack on a carbonyl carbon of an acid derivative, which gives a new acid derivative).

In this experiment, aniline is purified as its hydrochloride salt. Arylamines are relatively weak bases, but when treated with a strong mineral acid such as HCl, they are completely protonated to give the corresponding water-soluble hydrochloride salt:

$$Ph - NH_2 + HCl \rightarrow Ph - NH_3^+Cl^-$$

In this experiment, we use decolorizing charcoal, added to the aqueous solution of the hydrochloride salt, as a purifying agent. The charcoal adsorbs impurities, and the subsequent removal of the solution from the charcoal gives an aqueous solution of the purified hydrochloride salt.

The second part of the reaction sequence requires the sequential addition of acetic anhydride, followed by aqueous sodium acetate, to the reaction mixture. The purpose of the sodium acetate is to liberate the arylamine via an acid-base reaction so that the desired acyl substitution can occur. (Ammonium cations are not at all nucleophilic, as they are positively charged and do not possess a lone pair of electrons for nucleophilic attack.):

$$Ph - NH_3^+Cl^- + CH_3CO_2^- Na^+ \rightleftharpoons Ph - NH_2 + CH_3CO_2H + NaCl$$

Sodium acetate is the conjugate base of acetic acid, which is a relatively weak acid when compared to HCl. Furthermore, the anilinium ion, which has a pK_a of 4.6, is a slightly stronger acid than acetic acid, which has a pK_a of 4.8. Therefore, the equilibrium should shift to the right (towards the weaker acid), producing the desired arylamine.

PROCEDURE:

1. In the hood, place 100μL of aniline in a tared 10x75 mm (small) test tube. Place the test tube in a small beaker or Erlenmeyer flask to avoid spillage. Stopper the tube with a cork. Now, using a Pasteur pipet, add (with swirling) 0.5 mL of water, followed by three drops of concentrated HCl. Add 10 mg of decolorizing charcoal (we use the pelletized form, called Norit™) to the solution. Mix well, mindful not to crush the charcoal pellets.

2. Using a Pasteur pipet, transfer the solution to a 3-mL conical reaction vial, being careful not to take any of the pelletized charcoal along. Use an additional 0.5 mL of water to rinse the test tube and charcoal, and transfer that to the reaction vial as well. Place a magnetic spin vane in the vial and attach an air condenser.

3. Dissolve 150 mg of sodium acetate trihydrate in 0.5 mL of water in a 10x75 mm (small) test tube. Cork the tube and set aside in a small beaker or Erlenmeyer flask for use in a later step.

4. Remove the air condenser from the vial containing the anilinium chloride, and add 150 μL of acetic anhydride, followed quickly by addition of the previously prepared solution of sodium acetate. Reattach the air condenser and begin stirring. The reaction is very rapid, and the product should begin to precipitate almost immediately upon mixing of the reagents. Stir thoroughly to ensure proper mixing. Let the vial stand at room temperature for five minutes, and then place in an ice bath for an additional ten minutes to complete the crystallization process.

5. Collect the acetanilide by vacuum filtration using a Hirsch funnel. Rinse the conical vial with two 0.5 mL portions of cold water, and use the rinse to wash the collected solid. Pull air through the funnel for an additional ten minutes, and then allow the product to dry in your drawer in an open container until the next laboratory period.

6. Weigh the dry crystals and determine the melting point and percent yield. Recrystallization of the product is usually not necessary; however, if your melting point is more than 10 degrees off of the literature value, you *may* recrystallize the acetanilide from hot water (it will take approximately 3 mL of water to dissolve 150 mg of acetanilide).

Alkyl Halide Reactivities in Nucleophilic Substitution Reactions—Prelab

Prelab Report: Due at the Beginning of the Lab Period

Name ____

Lab Section (Circle One): Mon   Tues   Wed   Thur   Fri   AM/PM

1. The sodium iodide/acetone reagent classifies alkyl halides according to what?

2. The silver nitrate/ethanol reagent classifies alkyl halides according to what?

3. If a positive reaction is noted between NaI/acetone and 1-chlorobutane, what is the precipitate that is formed?

4. If a positive reaction is noted between silver nitrate/ethanol and t-butyl bromide, what is the precipitate that is formed?

5. What is a lachrymator?

6. Give the requested physical properties:

Boiling point of acetone:

Boiling point of ethanol:

Density of acetone:

Density of ethanol:

ALKYL HALIDE REACTIVITIES IN NUCLEOPHILIC SUBSTITUTION REACTIONS

INTRODUCTION

There are many factors that affect nucleophilic substitution reactions: substrate structure, nucleophile strength, leaving group ability, solvent, and temperature, among others. One of the most important of these is the substrate (alkyl halide) structure, because we find that the actual process (mechanism) of the substitution reaction changes depending on the structure of the alkyl halide reactant. The reaction can occur in a one-step process (where the nucleophile adds to the substrate simultaneously as the halide leaves):

$$\text{Nu:}^- \quad + \quad \text{R} - \text{X} \quad \longrightarrow \quad [\text{Nu} \text{-----} \text{R} \text{-----} \text{X}]^- \quad \longrightarrow \quad \text{Nu} - \text{R} \quad + \quad \text{X}^-$$

Or it can occur in a two-step process:

$$\text{R} - \text{X} \longrightarrow \quad \text{R}^+ \quad + \quad \text{X}^-$$

$$\text{Nu:}^- \quad + \quad \text{R}^+ \longrightarrow \quad \text{Nu} - \text{R}$$

These two reaction mechanisms have different reaction kinetics. The rate of the one-step process depends on the concentration of <u>both</u> the nucleophile and the substrate alkyl halide and is called a *bimolecular nucleophilic substitution reaction* (S_N2). The rate of the two-step process depends only on the concentration of the substrate alkyl halide and is called a *unimolecular nucleophilic substitution reaction* (S_N1).

Your lecture textbook will delve more deeply into these two reaction types. However, a typical order of reactivity for simple alkyl halides in an S_N2 reaction is as follows: methyl > primary > secondary > tertiary (although tertiary alkyl halides do not usually react via an S_N2 mechanism). The primary factor in this order of reactivity is steric hindrance (that is, the ease with which the nucleophile can approach the substrate carbon that bonds the halide). The typical order of reactivity for alkyl halides in an S_N1 reaction is thus: tertiary > secondary > primary > methyl (although methyl and primary alkyl halides do not usually react via an S_N1 mechanism). The primary factor in this order of reactivity is the relative stability of the carbocation that is formed in the first step of the two-step process (which is the rate-determining step). Allylic and benzylic carbocations are particularly easily formed, because they are resonance stabilized.

allylic cation resonance

benzylic cation resonance

Vinylic and aryl halides do not form carbocations, so they are usually unreactive in S_N1 reactions; the p-electrons from the adjacent double bond repel the nucleophile, so they are usually unreactive in S_N2 reactions.

aryl halide   vinyl halide

We will utilize two classical sets of reaction conditions:

1. *Sodium iodide in acetone*, a system which uses the iodide ion as the nucleophile and acetone as a relatively nonpolar aprotic solvent. These reaction conditions favor an S_N2 reaction. Sodium iodide is soluble in acetone, but both sodium bromide and sodium chloride are not soluble. Therefore, it is easy to identify if a substitution reaction occurs (the formation of the sodium halide precipitate indicates a reaction has occurred):

$$Na^+I^- \; + \; R-Br \, (or \; R-Cl) \; \longrightarrow \; R-I \; + \; NaBr\downarrow (or \; NaCl\downarrow)$$

2. *Silver nitrate in ethanol*, a system which contains the very poor nucleophile nitrate ion and ethanol as a polar protic (and nucleophilic) solvent. These reaction conditions favor an S_N1 reaction. The silver ion greatly assists the ionization of the halide ion by coordinating with an electron pair of the halogen as the insoluble silver halide salt forms. Again, it is easy to identify if a substitution reaction occurs (the formation of the silver halide precipitate indicates a reaction has occurred):

$$R-X \; + \; Ag^+ \; \longrightarrow R-XAg \longrightarrow R^+ \; + \; AgX\downarrow$$

EXPERIMENTAL PROCEDURE

Number a series of clean and dry 10 × 75 mm test tubes from 1 to 11. Each test tube will be used for one of these organic halides [halide (1) in test tube 1; halide (2) in test tube 2; etc]:

Test Tube #	Compound	Test Tube #	Compound
1	1-bromobutane	6	2-chlorobutane
2	1-chlorobutane	7	2-chloro-2-methylpropane
3	bromocyclopentane	8	benzyl chloride
4	2-bromobutane	9	1-chloro-2-butene [crotyl chloride]
5	bromocyclohexane	10	bromobenzene

1. Place 2 mL of a 15% sodium iodide in acetone solution into each test tube; then add four drops of the appropriate halide to each of the test tubes. Stopper each test tube, shake the mixture,[1] and record the time. Now note the time needed for any reaction to occur (cloudiness or precipitate). If no reaction occurs after five minutes, place only those test tubes (with no solid) into a 50°C water bath.[2] Again note the time needed for any reaction to occur (at the elevated temperature). After five minutes of heating, remove the test tubes from the water bath, and record which of the halides have remained unreactive. Empty the test tubes into the halogenated organic waste container.[3] Continue to part two.

2. Place 2 mL of a 1% ethanolic silver nitrate solution into each test tube. Then add four drops of the appropriate halide to each of the test tubes. Stopper each test tube, shake the mixture,[1] and record the time. Now note the time needed for any reaction to occur (cloudiness or precipitate). If no reaction occurs after five minutes, place only those test tubes (with no solid) into a 50°C water bath. Again note the time needed for any reaction to occur (at the elevated temperature). After five minutes of heating, remove the test tubes from the water bath, and record which of the halides have remained unreactive. Empty the test tubes into the halogenated organic-waste container. Continue to part three.

3. Rinse test tube number 6 with ethanol, and allow it to drain (it does not need to be dry). Repeat the ethanolic silver nitrate reaction (as in part two) using 1% silver nitrate in a 50% ethanol–50% water solvent. The ethanol-water solvent is a more polar solvent (compared to ethanol alone).

OBSERVATIONS

You are looking for solid formation to indicate that a reaction has occurred (so if there is a color change but no cloudiness or precipitate, the color change should be recorded, but there is no reaction). Other qualitative observations might include the following: cloudy, slight ppt, dense ppt, green liquid + pale yellow ppt, etc. Be as complete as possible.

REPORT

In the conclusion of your report, you should address these variables (in both S_N1 and S_N2 reaction conditions):

1. Substrate structure (1°, 2°, 3°, allylic, aromatic, benzylic, and vinylic)

2. Reaction temperature

3. Solvent polarity

Your discussion should do the following:

1. Compare and comment on the reactions of 2-bromobutane and 2-chlorobutane

2. Compare and comment on the reactions of 2-chlorobutane with regard to solvent polarity

3. Compare and comment on the reactions of benzyl chloride and 1-bromobutane (both are 1°);

4. Compare and comment on the reactions of benzyl chloride and bromobenzene (both are aromatic)

QUESTION

If you were given these three compounds, how would you expect them to react under both S_N1 and S_N2 reaction conditions? Explain your answer.

---

[1] Your instructor may prefer an alternate way of mixing the contents of the test tube (by "flicking" the bottom of the tube).

[2] You must regulate the temperature of the hot-water bath carefully (acetone boils at 56.5°C).

[3] Your instructor will tell you what to do with the test tubes. They can be washed well, rinsed with ethanol, and allowed to dry.

Benzoin Condensation—Prelab

Prelab Report: Due at the Beginning of the Lab Period

Name ____

Lab Section (Circle One): Mon   Tues   Wed   Thur   Fri   .   AM/PM

1. Define a chemical catalyst.

2. Why is a sodium hydroxide solution added to the thiamine solution?

3. What structural features do cyanide and thiamine have in common that makes them both capable of catalyzing the benzoin condensation?

4. If you started with 15 mL of benzaldehyde and sufficient quantities of all other reagents, and obtain benzoin in a 70% yield, what is the mass <u>in grams</u> of your product? (Show all your calculations.)

5. Which of the "B" vitamins is thiamine?

6. Provide the requested physical properties:

Melting point, benzaldehyde:

Moleculat weight, thiamine HCl:

Density, benzaldehyde:

Melting point, benzoin:

Color of benzoin:

Benzoin Condensation— A Microwave Experiment

INTRODUCTION

The classic benzoin condensation is the catalytic dimerization of benzaldehyde to give 2-hydroxy-1,2-diphenylethanone (benzoin). From a green-chemistry perspective, this reaction is very efficient, exhibiting complete atom economy, as all atoms from the benzaldehyde are retained in the final product.

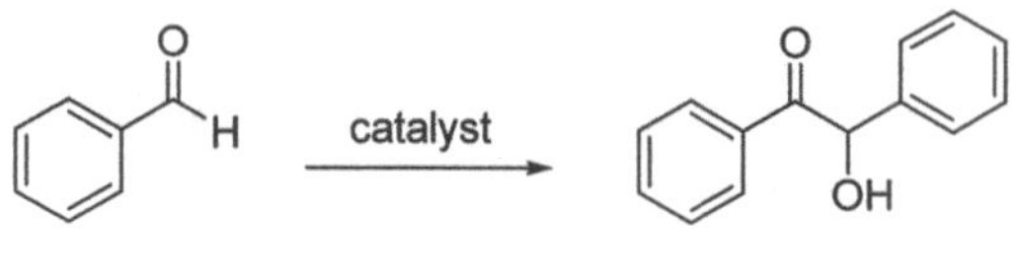

Scheme 1—*Dimerization of benzaldehyde*

This is not a spontaneous reaction: a catalyst is required. Initially, the catalyst attacks the carbonyl carbon, giving an intermediate that rearranges to give a benzylic anion. This anion then acts as a nucleophile, attacking the carbonyl carbon of a second benzaldehyde molecule. The first version of this reaction used a cyanide catalyst, as shown below. Cyanide is extremely poisonous, however, and its use should be avoided if a safer alternative is available.

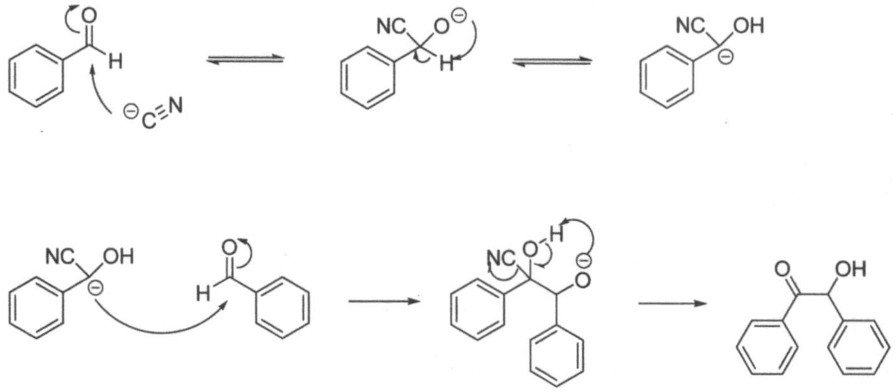

Scheme 2—*Mechanism of the cyanide-catalyzed benzoin condensation*

Similar sorts of dimerization and condensation processes are very common in biological systems. One of the major biochemical catalysts is the coenzyme thiamine pyrophosphate, also called vitamin B1.

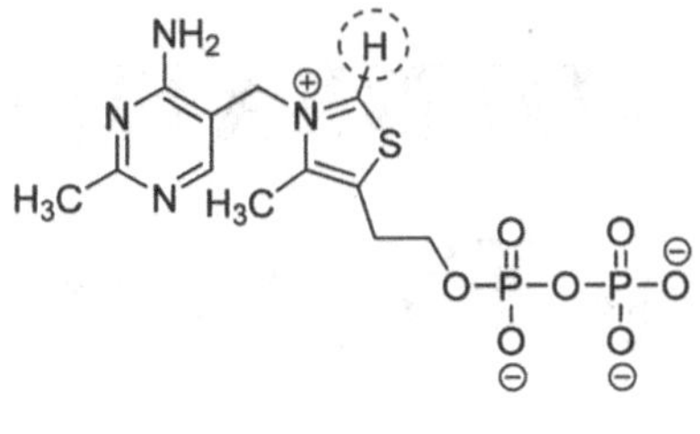

Figure 1—*Thiamine pyrophosphate*

The indicated hydrogen (on the 2-position of the thiazolidine ring) is very acidic. In biological systems, this hydrogen is easily removed to provide the anion, which then adds to pyruvate. Subsequently, carbon dioxide is lost, giving an acetyl derivative of thiamine pyrophosphate. There are many reaction pathways available in physiological systems at this point; one option is the reaction of the acetyl derivative with acetaldehyde, which produces acetoin, the acetaldehyde equivalent of benzoin.

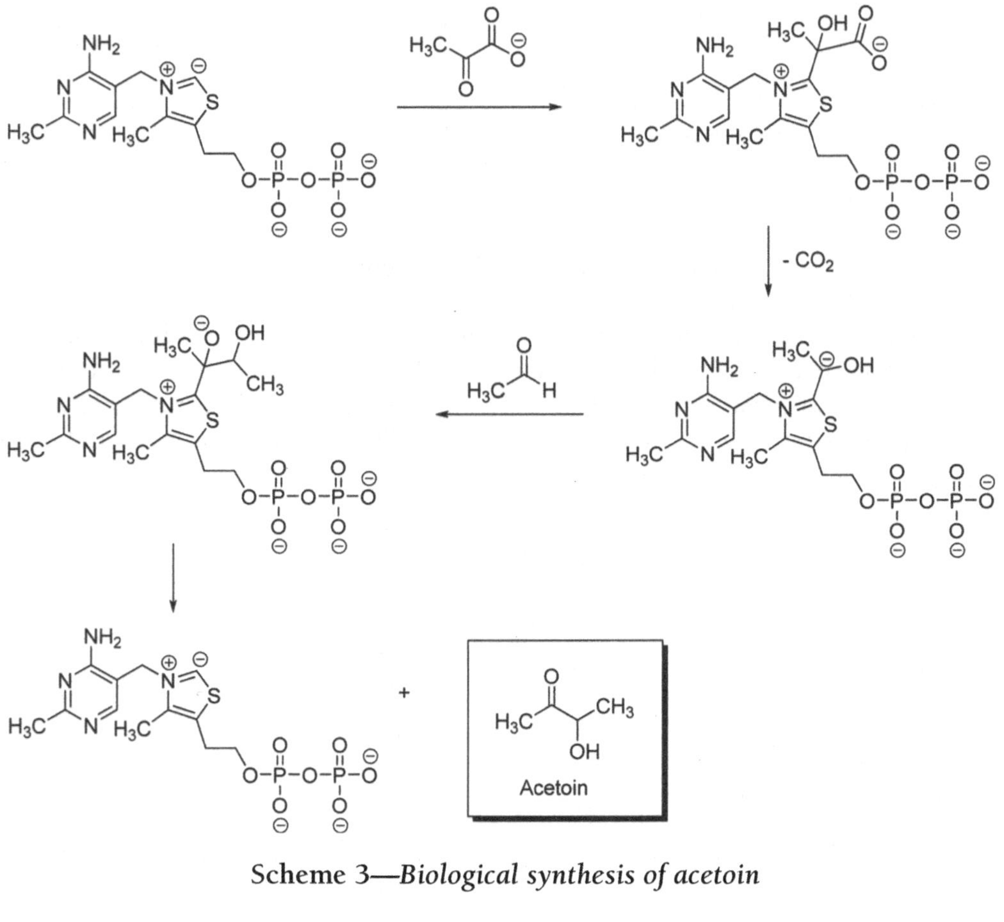

Scheme 3—*Biological synthesis of acetoin*

The use of thiamine hydrochloride (without the pyrophosphate appendage) as a catalyst in place of the highly toxic cyanide anion is an excellent illustration of alternative reagent selection following the precepts of green chemistry. Thiamine is edible, nontoxic, and significantly "greener" than cyanide.

The standard benzoin condensation using the thiamine catalyst requires a ninety-minute reflux, often involves a somewhat difficult isolation of product (as the product can be produced as an oil rather than as a solid), and does not produce particularly good yields. We have improved the procedure considerably by utilizing microwave chemistry, shortening the reaction time considerably, and giving the product in much better yield.

Because of its improved rate of reaction, microwave irradiation is considered a green laboratory technique.

PROCEDURE

1. Dissolve 0.35 g of thiamine hydrochloride in 1 mL of water in a microwave GlassChem vessel. Add 3 mL of 95% ethanol, and cool the resulting solution by swirling the vessel in an ice-water bath. Separately, place 1 mL of 2M NaOH in a small test tube, and cool in an ice bath. After about five minutes of cooling, slowly add the cold sodium hydroxide solution dropwise to the thiamine hydrochloride solution, using a Pasteur pipet. Continue the NaOH addition until the solution turns a clear, yellow color or until all of the NaOH solution has been added. Then, add 2 mL of benzaldehyde to the thiamine-NaOH mixture. Add a stir bar, and cap the microwave vessel. Place the microwave tube into the turntable (be sure to record the number of the slot), and secure it.

2. Your instructor will place the turntable into the MARS microwave reactor when all of the tubes are secured. The microwave reactor will then be set to 80°C and twenty minutes of reaction time.

3. When the microwave reaction cycle is complete, remove your tube from the turntable, and stand it in a beaker for easy transport. The solution should be clear and homogeneous; if an oily layer forms, ask your instructor for assistance. Let the solution cool to room temperature, and then place the tube into an ice bath to induce crystallization. Scratch the sides of the tube with a glass rod, if needed, to initiate the crystallization process.

4. After crystallization is complete (ten to twenty minutes in the ice bath), collect the crude product by vacuum filtration. Wash the solid product twice with 5 mL of ice-cold water. Recrystallize your crude product from 95% ethanol (the solubility of benzoin is about 12 g/100 mL ethanol). Allow your product to dry. In the next laboratory period, weigh your product, calculate the percent yield, determine the melting point of your product, and obtain an IR of your product, as well as an IR of benzaldehyde for comparison.

Benzoin Condensation—Alternate (Non-Microwave) Procedure

PLEASE NOTE: WORK IN GROUPS OF 2-3 IF USING THIS PROCEDURE

Dissolve 3.5 g of thiamine hydrochloride in about 10 mL of water in a 250-mL Erlenmeyer flask. Add 30 mL of 95% ethanol and cool the resulting solution by swirling the flask in an ice-water bath. Meanwhile, place about 10 mL of 2M sodium hydroxide solution in a small Erlenmeyer flask, and also cool this solution in the ice bath. After about five minutes of cooling, slowly (over the next ten minutes) add the cold sodium hydroxide solution to the thiamine solution. Add 20 mL of benzaldehyde to the thiamine mixture. Add a boiling stone and heat the resulting mixture gently on a hot plate for about 75–90 minutes. DO NOT boil the mixture vigorously! Allow the mixture to cool to room temperature, and then induce crystallization of benzoin (it may have already started) by cooling the mixture in an ice-water bath. If the product separates as an oil, reheat the mixture until it once again is homogeneous, and then allow it to cool more slowly. You may need to scratch the inside wall of the flask with a glass rod to induce crystallization.

After crystallization is complete (about 20 minutes) collect the crude product by vacuum filtration using a Büchner funnel. Wash the solid product twice with 50 mL portions of ice-cold water. Recrystallize your crude product from 95% ethanol (the solubility of benzoin is about 12 g/100 mL ethanol). Allow your product to dry. Next lab period, weigh your product, calculate the percent yield, determine the melting point of your product, and obtain an IR of your product as well as an IR of benzaldehyde for comparison.

SYNTHETIC REACTION SEQUENCE

You will continue a four-step reaction sequence as outlined below. This sequence will involve the oxidation of benzoin to benzil, then use of the benzil in an aldol condensation to prepare tetraphenylcyclopentadienone, the required diene for the final step: a Diels-Alder reaction with diphenylacetylene to prepare hexaphenylbenzene.

1. 2 eq.

Benzaldehyde

Benzoin

2.

Benzoin

Benzil

3.

Benzil + 1,3-Diphenylacetone → Tetraphenylcyclopentadienone

Tetraphenylcyclopentadienone + Diphenylacetylene ⇌ → Hexaphenylbenzene

BROMINATION OF AN ALKENE—PRELAB

Name ____

Lab Section (Circle One): Mon   Tues   Wed   Thurs   Fri   AM/PM

1. What is the limiting reagent for this reaction?

2. Give a mechanism for the reaction, assuming that Br_2 is the electrophile.

3. What color do you expect the solution to be at the end of the reaction?

4. Provide the requested physical properties:

    a. melting point, E-stilbene

    b. molecular weight, pyridinium tribromide

    c. density, ethanol

    d. boiling point, ethanol

BROMINATION OF AN ALKENE[1]

BACKGROUND

A common method for adding functionality to simple hydrocarbons (alkanes and alkenes) involves the addition of bromine to a double bond to form a vicinal dibromide. This is now capable of many new transformations, allowing elaboration of the original molecule into something much more complex.

Addition of bromine to an alkene is generally considered to follow an ionic mechanism, in which the alkene acts as a nucleophile and the bromine molecule acts as an electrophile. (In general, electron-rich species make good nucleophiles, and electron-deficient species make good electrophiles.) We do not necessarily think of molecular bromine as electron deficient, but as a bromine molecule approaches the electron-dense pi system of the alkene, the bromine-bromine bond becomes polarized, as shown below.

$$\text{Br}-\text{Br} \longrightarrow \overset{\delta^+\ \ \delta^-}{\text{Br}-\text{Br}}$$

In this experiment you will perform a bromination of *E*-stilbene (*trans*-stilbene). Historically, bromination of an alkene was done using molecular bromine in a chlorinated solvent such as methylene chloride. However, not only is molecular bromine volatile and highly corrosive (in addition to causing severe burns upon contact with skin and irritation upon inhalation), but methylene chloride is a suspected carcinogen. A series of modifications have been made to this historical procedure, each of which has improved upon the original reaction. One modification is the solvent. Instead of methylene chlorine, many researchers chose to use glacial acetic acid. Unfortunately, while not carcinogenic, glacial acetic acid is also somewhat volatile and quite corrosive. You will use ethanol as the solvent, a much safer alternative.

An attractive substitute to molecular bromine is pyridinium tribromide (originally reported by Djerassi and Scholz[2]), a solid that, in solution, exists in a rapid equilibrium with pyridinium hydrobromide and molecular bromine, as shown below. This allows the production of small amounts of bromine *in situ*, avoiding the inherent problems with use of molecular bromine as the actual reagent.

---

[1] Based on Doxsee, Kenneth M., and James E. Hutchison. "Bromination of an Alkene: Preparation of Stilbene Dibromide." In *Grenn Organic Chemistry: Strategies, Tools, and Laboratory Experiments*. Brooks/Cole, 2004. 120–124.
[2] C. Djerassi and C.R. Scholz, *J. Am. Chem. Soc.* **1948**, 70, 417.

PROCEDURE

CAUTION: Use caution when handling the materials used in this experiment. Wear gloves and appropriate eye protection at all times during this procedure. Pyridinium tribromide is corrosive and a lachrymator. Clean up any spills immediately, especially on the balance, the metal parts of which will quickly be corroded.

Place 0.200 g of *E*-stilbene and 2 mL solvent-grade (95%) ethanol in a 10-mL roundbottom flask with a stir bar. Add 0.400 g of pyridinium tribromide to the solution, using an additional 2 mL of ethanol as necessary to rinse down the sides. Attach an air condenser and clamp the setup in place over a stirrer/hot plate using a ring stand.

Reflux with stirring for five minutes. The color of the solution should turn from red to yellow during this time period. Once the reaction is complete, remove from the heat (just turn the clamp away from the hot plate) and allow to cool to room temperature. The product should begin to crystallize as the solution cools.

Once the solution is at room temperature, transfer the roundbottom flask to an ice bath and continue cooling for ten minutes. Collect the product by vacuum filtration. Wash the isolated solid with a small amount of ice-cold ethanol to remove any adsorbed pyridinium salts.

Once your product is dry (next laboratory period), weigh it, obtain a melting point, and collect an IR spectrum. (Ask your TA for help if necessary.)

QUESTIONS

1. What are the possible stereoisomeric products?

2. Predict the most likely product and explain your reasoning.

3. What product was actually produced? How do you know this?

4. Calculate the atom economy for the reaction. How does this compare to an alternative procedure using [$HBr + H_2O_2 \rightarrow Br_2 + 2H_2O$] as the reagent?

LAB REPORT

Include the weight, melting point, and percent yield calculation. Attach the IR of your product with a basic analysis. Answer the questions above.

DISTILLATION EXPERIMENT— PRELAB

Prelab Report: Due at the Beginning of the Lab Period

Name ____

Lab Section (Circle One):   Mon   Tues   Wed   Thur   Fri   AM/PM

1. What piece(s) of glassware distinguish a simple distillation setup from a fractional distillation setup?

2. Under what conditions can a good separation be achevied using a simple distillation?

3. Define a theoretical plate.

4. What is the mole percent of methanol in your starting solution (composed of 1.5 mL methanol and 6 mL isopropanol)?

5. Using your vapor-liquid equilibrium data graph (NOTE: You have to make this graph using the data provided in the experimental procedure!) and your answer from question #4, answer the following:

a. If your first distillate was 49 mole percent methanol, how many theoretical plates were in your distillation setup?

b. If your first distillate was 87 mole percent methanol, how many theoretical plates were in your distillation setup?

6. Provide the requested physical properties:

Boiling point of methanol: Refractive index of methanol:

Boiling point of isopropanol: Density of isopropanol:

DISTILLATION EXPERIMENT

PROCEDURE

Simple Distillation

1. A mixture of 1.5 mL of methanol and 6 mL of isopropanol is placed in a 10-mL round-bottom flask with two or three boiling chips. A small sample (approximately 0.5 mL) of the original mixture should be retained in a capped vial for analysis. The mixture is distilled using a simple distillation apparatus (see illustration below). The first 0.5 mL of distillate is collected in a vial, capped, and saved for analysis; the rest of the distillate is collected in a 10-mL graduated cylinder. Make a data table in your laboratory notebook; record the temperature after every 0.5 mL of distillate. This data will then be graphed.

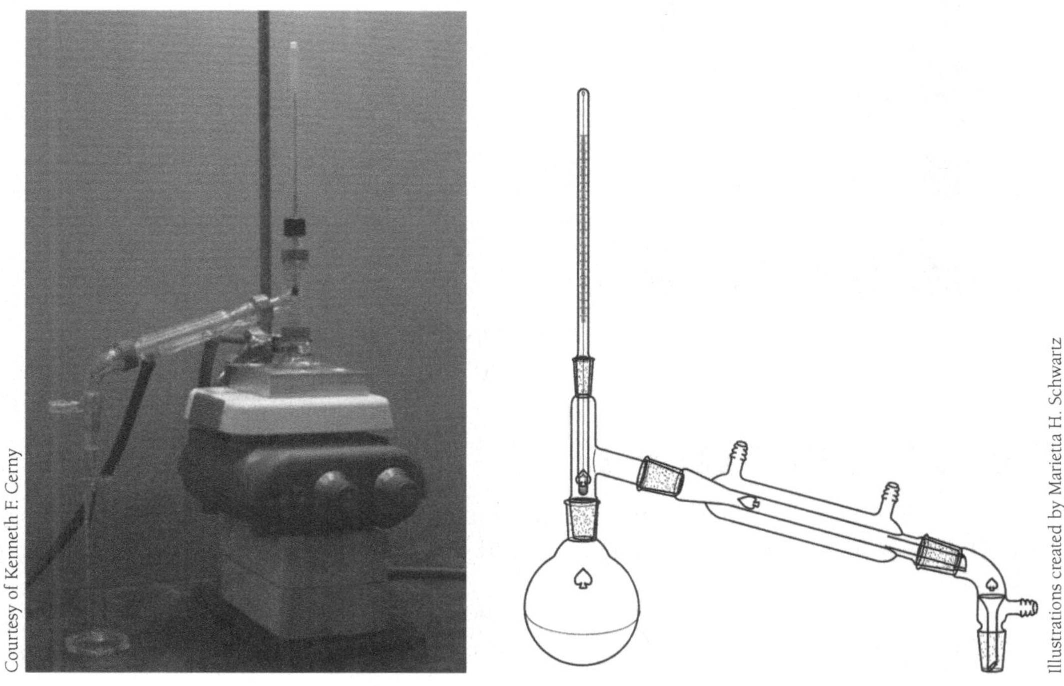

Figure 1—*Simple Distillation Setup*

Fractional Distillation

1. A second mixture of 1.5 mL of methanol and 6 mL of isopropanol is placed in a 10-mL round-bottom flask with two or three boiling chips. A small sample (approximately 0.5 mL) of the original mixture should be retained in a capped vial for analysis. The mixture is distilled using a fractional distillation apparatus (see illustration below; this is essentially a simple distillation with a fractionating column added). The first 0.5 mL of distillate is again collected in a vial, capped, and saved for analysis; the rest of the distillate is collected in a 10-mL graduated cylinder. Make a data table in your laboratory notebook; record the temperature after every 0.5 mL of distillate. This data will then be plotted on the same set of axes as the data for the simple distillation.

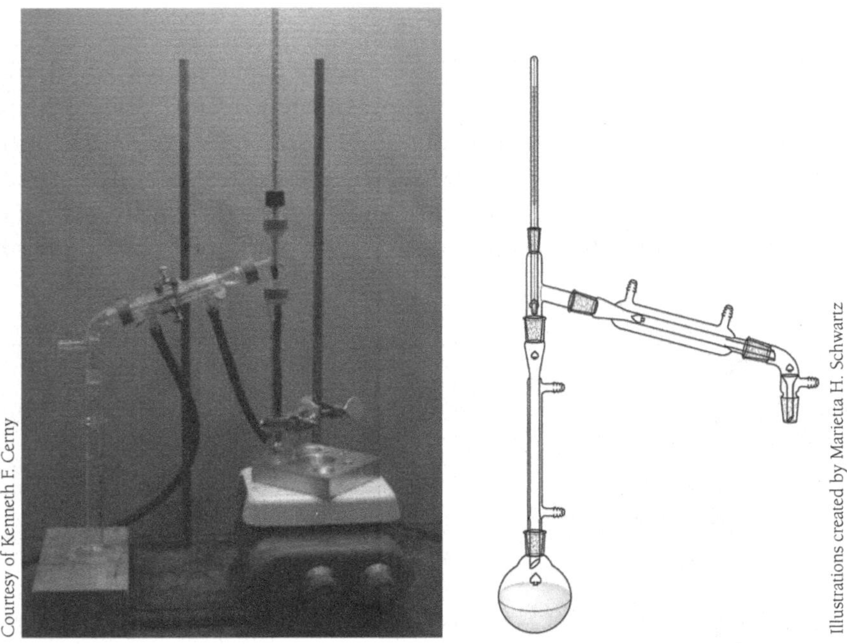

Figure 2—Fractional Distillation Setup

A NOTE ON GRAPHING

You will be generating three (or four, depending on how you count it) graphs for this experiment. <u>Each graph should take up a full page in your laboratory notebook, with the axes set up so as to spread out the data points as much as possible</u>. The graphs are as follows:

1. **Vapor–Liquid Equilibrium Data:** (This should be done <u>before the experiment</u>, as you need this graph to answer some of the prelab questions). This graph is generated from the table of data provided on page 99 as Table 1.

2. **Temperature vs. Volume of Distillate:** Put the data for simple and fractional distillations on the same set of axes.

3. **Refractive Index vs. Volume Percent:** Use your own measurements of refractive indexes for pure methanol and pure isopropanol, not the literature values.

DATA ANALYSIS

1. Compare the graphs you drew for the simple and fractional distillation. What does this tell you about the relative efficiency of the two techniques? <u>Discuss this in your laboratory report</u>.

2. Analyze the four samples you have collected (two original mixtures and two first distillates) by refractive index, along with the pure components (pure methanol and pure isopropanol). Compare the efficiency of the two techniques by calculating the number of theoretical plates in each apparatus. <u>Your calculations and conclusions should also be in your laboratory report (and notebook)</u>.

Analysis by Refractive Index

1. Measure the refractive index of the four samples and the two pure components, using the refractometers available in the laboratory. Record each refractive index (abbreviated as "n_D") to four decimal places.

2. Assuming that the refractive index is a linear function of the composition of the sample, the volume percent of the four experimental samples can be calculated:

$$n_D(1 + 2) = \left[\frac{\text{volume}_1}{\text{total volume}} \times n_D(1) \right] + \left[\frac{\text{volume}_2}{\text{total volume}} \times n_D(2) \right]$$

3. The simplest method to perform this calculation involves preparation of graph 3 (mentioned above). See Figure 3 for an example. The refractive index is plotted on the y-axis, and the volume percent of one of the components is plotted on the x-axis. Plot the refractive index of pure methanol (as measured by you) on one side (A) and of pure isopropanol on the other side (B). Two points determine a line; connect the two points.

4. Then find the refractive index for one of your samples (C). Read over to the straight line and then down, and you will have the volume-percent composition of your sample. This value can be converted to weight-percent composition and then to mole-percent composition, using the appropriate conversion factors; see the sample calculations below.

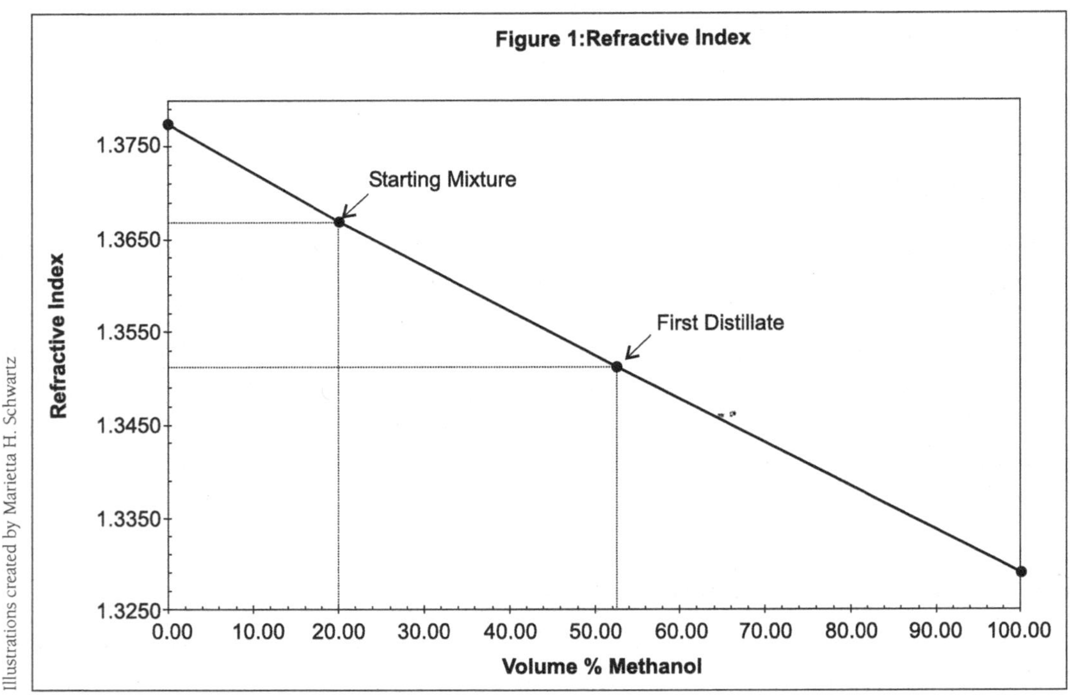

Figure 3—Refractive Index

Sample Calculations

1. For a simple distillation, the refractive index of the starting mixture was 1.3669, and that of the first distillate was 1.3511. Using the graph from Figure 3 to convert the refractive indices to volume percent gives 19.7 volume percent methanol for the initial mixture and 52.1 volume percent methanol for the first distillate.

2. To utilize the vapor-liquid equilibrium graph (Figure 4) to convert percentages to theoretical plates, the volume percentages must first be converted to mole percentages. This is done assuming 100 mL total volume of solution and using the densities and molecular weights of the two components. Following that logic, 19.7 volume percent methanol becomes 19.7 mL methanol, leaving the rest of the sample to consist of 80.3 mL isopropanol:

$$19.7 \text{ mL MeoH} \times \frac{0.791 \text{g}}{\text{mL MeoH}} \times \frac{1 \text{ mol MeoH}}{32.04 \text{ g}} = 0.486 \text{ mol MeoH}$$

$$80.3 \text{ mL iProH} \times \frac{0.785 \text{g}}{\text{mL iProH}} \times \frac{1 \text{ mol iProH}}{60.10 \text{ g}} = 1.049 \text{ mol iProH}$$

Once the number of moles of each component has been calculated, you can then calculate the mole percent of methanol:

$$\text{Mole \% MeoH} = \frac{\text{Moles MeOH}}{\text{Total Moles}} = \frac{0.486}{(0.486 + 1.049)} \times 100\% = 31.7 \text{ mole \% MeOH}$$

Sample	Refractive Index	Volume %	Mole %
Starting Mixture	1.3669	19.7	31.7
First Distillate	1.3511	52.1	67.3

Theoretical-Plate Calculation

1. Using the mole-percent composition of the starting mixture and initial distillate for each type of distillation, the number of theoretical plates can be calculated. Basically, a theoretical plate corresponds to a vaporization-condensation cycle. The more of these cycles that occur during a distillation, the more efficient the separation of the two components will be. The number of theoretical plates for a particular distillation can be calculated using a vapor-liquid equilibrium data graph. You can construct such a graph (see Figure 4) using the data provided in Table 1.

Table 1 Experimental Vapor-Liquid Equilibrium Data for Isopropanol/Methanol System at 760 mm Hg

		Mole % of Methanol	
Temperature (°C)	Vapor	Liquid	
66.22	95.35	90.10	
67.94	89.10	79.00	
70.22	80.00	66.05	
72.67	68.50	52.20	
74.78	57.00	40.80	
77.06	42.85	29.30	
78.94	29.60	19.50	
81.00	13.20	8.10	

Source: Data from L.H. Ballard and M. Van Winkle, "Vapor-Liquid Equilibrium at 760 Mm. Pressure. 2-Propanol-Methanol, 2-Propanol-Ethyl Alcohol, 2-Propanol-Propanol, and 2-Propanol-2-Butyl Alcohol Systems," Industrial and Engineering Chemistry 44, no. 10 (1952): 2450–2453.

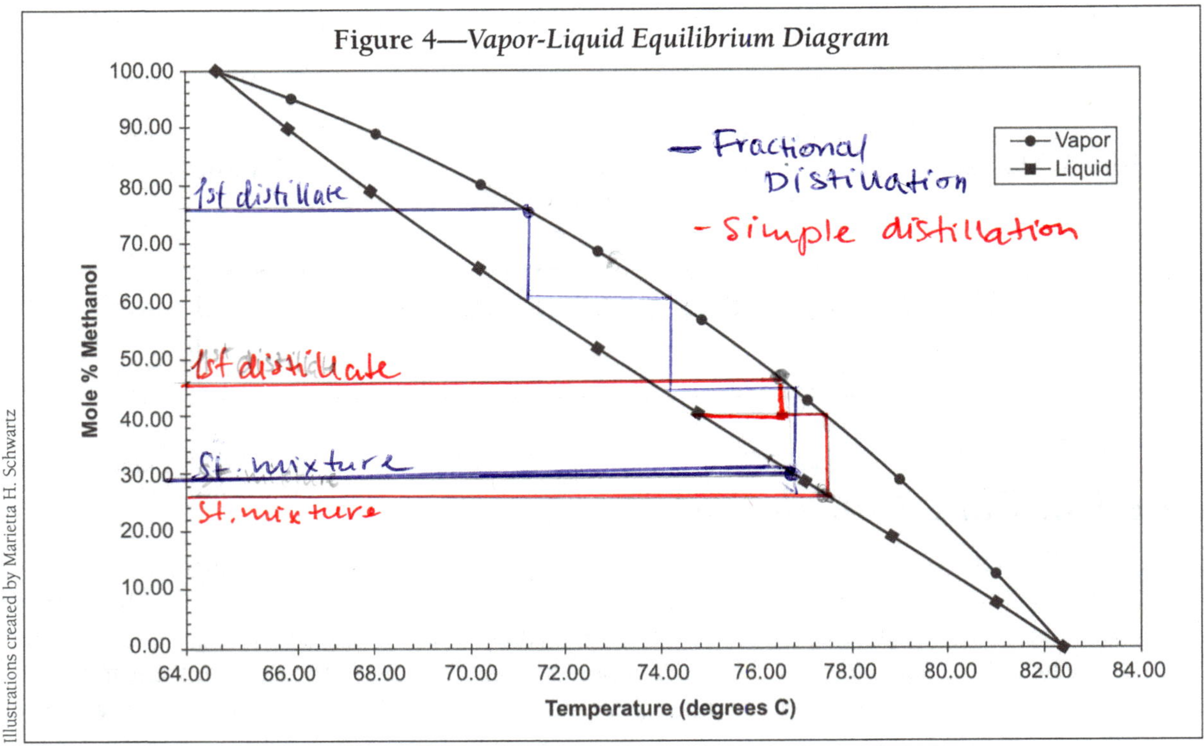

Illustrations created by Marietta H. Schwartz

2. To calculate theoretical plates, find the mole-percent composition of your starting mixture on the liquid line. "Vaporize" your sample by drawing a line straight up until it connects with the vapor line (note that there is no temperature change). Then "condense" your sample by drawing a line straight over to the left (cooling temperatures) until it connects with the liquid line again. This "stair step" is a vaporization-condensation cycle, commonly known as a theoretical plate.

3 Repeat this process until you get to the point on the liquid line corresponding to the mole-percent composition of your initial distillate sample. The number of stair steps is the number of theoretical plates.

4. If you use the data from the sample calculation above, you will see that the simple distillation used as an example contained approximately two theoretical plates.

THE SYNTHESIS OF HEXAPHENYLBENZENE—PRELAB

Prelab Report: Due at the Beginning of the Lab Period

Name ____

Lab Section (Circle One): Mon   Tues   Wed   Thur   Fri   AM/PM

1. This preparation is an example of what name reaction?

2. What is the structure of the dienophile?

3. Why is the initially formed bicyclic adduct not isolated?

4. What is the purpose of the hexane and toluene that are added?

5. What starting materials would you use to prepare the compound shown below by this same type of reaction?

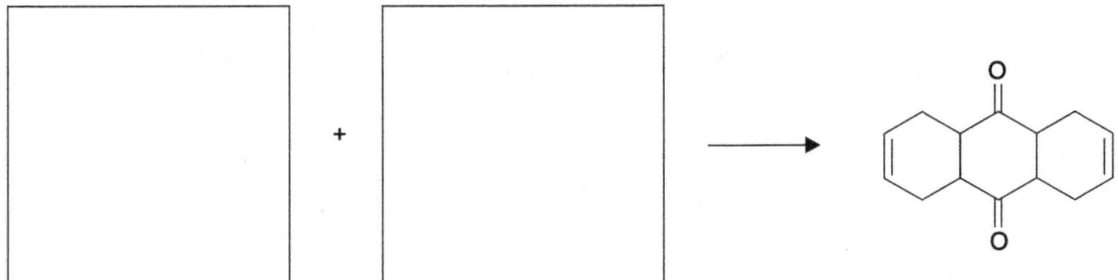

6. Provide the requested physical properties:

Melting point, diphenylacetylene:

Melting point, tetraphenylcyclopentadienone:

Boiling point, hexane:

Boiling point, toluene:

THE SYNTHESIS OF HEXAPHENYLBENZENE

INTRODUCTION

The molecule to be prepared in the Sequence A synthesis is hexaphenylbenzene (I).

I

Hexaphenylbenzene

This system, which contains seven aromatic rings, was first made by Dilthey at the University of Bonn in 1933 by employing a classic Diels-Alder reaction with exactly the same two reactants as you will generate in Sequence A. The Bonn group showed that an earlier claim, by Durand, to have prepared this compound via a massive Grignard attack by phenylmagnesium bromide on hexachlorobenzene, had not actually yielded hexaphenylbenzene. The compound isolated by Durand melted at 266 °C, while Dilthey's material melted at 421–422 °C. Hexaphenylbenzene was later synthesized photochemically by Büchi at the Massachusetts Institute of Technology (MIT) in 1962. The MIT group improved the purity of the isolated material, and reported a melting point of 439–441 °C. Fieser, at Harvard University, refined Dilthey's synthetic route, and published the definitive preparation in Organic Syntheses in 1966. Fieser obtained melting points without decomposition in evacuated melting point capillaries in the range 454–456 °C.

This hydrocarbon system possesses a number of interesting structural and physical properties. First, we should note that it has a relatively high molecular weight, near 534, and a molecular formula of $C_{42}H_{30}$. Hexaphenylbenzene exhibits an extremely high melting point for a nonionic organic material. For example, of the 15,000 plus substances listed in the Table of Physical Constants for Organic Compounds in the CRC Handbook of Physics and Chemistry, only two materials melt above 450 °C (and both of these compounds decompose at their melting point), and only 10 melt above 400 °C. Indeed, hexaphenylbenzene melts above hexabenzobenzene (II) the completely delocalized seven-ring fused system, mp = 438–440 °C.

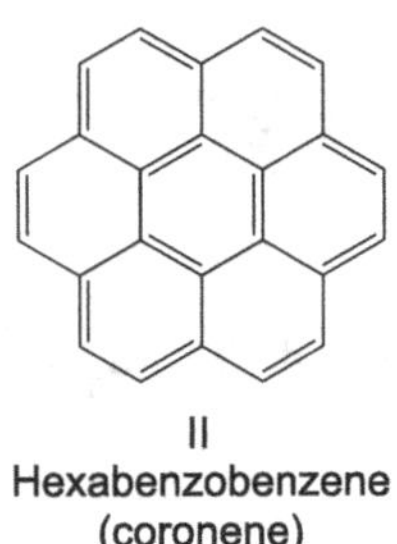

II
Hexabenzobenzene
(coronene)

From molecular modeling studies of hexaphenylbenzene, it appears that the six substituent rings surrounding the central system will be sterically restricted from lying in the plane of that ring by ortho-position interaction (III)

III
Hexaphenylbenzene with steric interactions
Between ortho positions

and the figure below shows a molecular model of one of the chiral rotamers of hexaphenylbenzene.

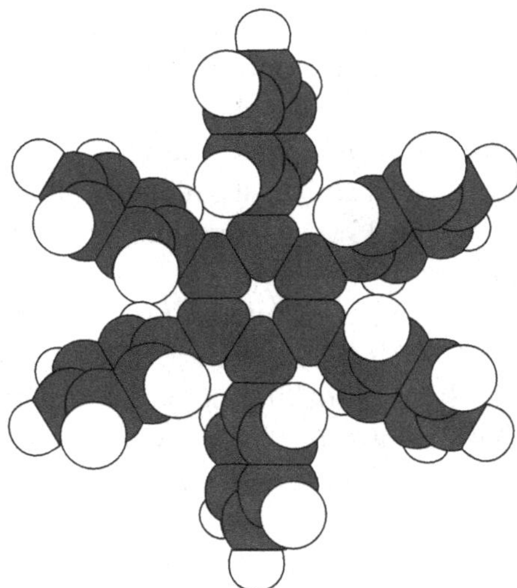

Figure 1—*Molecular model of one of the chiral rotamers of hexaphenylbenzene.*

The twisted structure presents a particularly interesting problem in stereoisomerism. It is clear that when all the rings are coplanar (dihedral angle of 0°), we have maximum overlap of the p system and delocalization energy, but we also have a maximum of steric repulsion energies. On the other hand, when the dihedral angle approaches 90°, all delocalization is blocked, though steric repulsion between rings is at a minimum. It would seen reasonable to expect the system to reach some energetic compromise between these two extreme orientations. If this is the case, is it possible to establish the angle at which the external rings are positioned? An X-ray crystallographic study carried out by Bart in 1968 on solid crystalline hexaphenylbenzene did, in fact, show that in the crystal lattice the phenyl groups are twisted 65° out of the plane of the central ring. In the crystal lattice, the molecule adopts a conformation, similar to a six-bladed propeller, which is chiral. That is, in the solid state hexaphenylbenzene can exist in two enantiomeric forms. Indeed, if this molecule happened to undergo resolution of the optical isomers during crystallization, in much the same fashion as Pasteur's tartaric acid salts, it should be possible to mechanically separate the racemate into crystals in which all the rings are tilted only in one possible conformation, or in the other. These enantiomers, however, should possess a relatively low barrier to rotation, so that when dissolved in solution rapid racemization, via rotation of the rings (propeller blades) to the opposite pitch, would be expected.

This does seem to be the case. Gust at Arizona State University showed in 1977 that if a derivative of hexaphenylbenzene were synthesized in which two different groups were substituted on adjacent rings in the ortho positions (e.g., a methyl group and a methoxy group, IV), two possible sets of diastereomers would result.

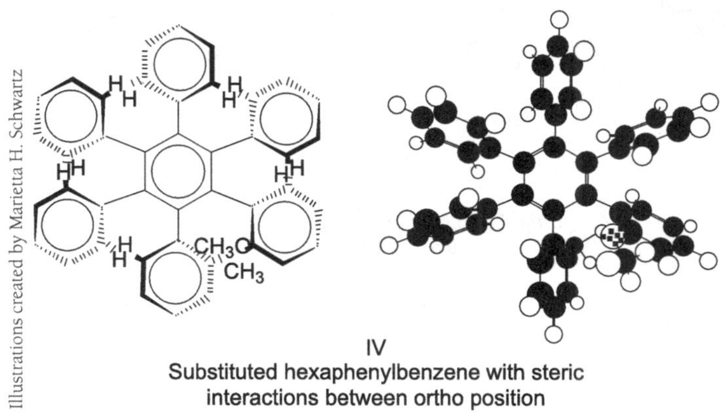

IV
Substituted hexaphenylbenzene with steric
interactions between ortho position

It is presumed that it would be impossible for the rings to rotate the two ortho substituents past one another; but that other rotations may or may not be facile. If a large rotational barrier were present, the external rings would remain tilted at 65° with the same pitch. If this were the case, we would expect four diastereomers, and thus four different C–CH$_3$ resonances in the ^{1}H NMR spectrum. If rapid interconversion of the tilted conformers occurred, we would expect that the two bulky ortho groups would restrict full rotation of those two rings, even though the barrier to pitch inversion is low. Thus, in this latter case we would expect two diastereomeric pairs of enantiomers (one with the two ortho groups up and one with one up and one down), and two different C–CH$_3$ resonances in the NMR. When the compound was synthesized, and its ^{1}H NMR spectrum obtained, two resonances for methyl groups attached to aromatic rings, and two O–CH$_3$ resonances were observed. These two diastereomers, V, were separated by column chromatography, and it was found that they slowly interconverted upon being heated to 215 °C. Thus, on the NMR time scale, it would appear that in hexaphenylbenzene, a reasonably rapid interconversion of the propeller conformations is taking place in solution at room temperature.

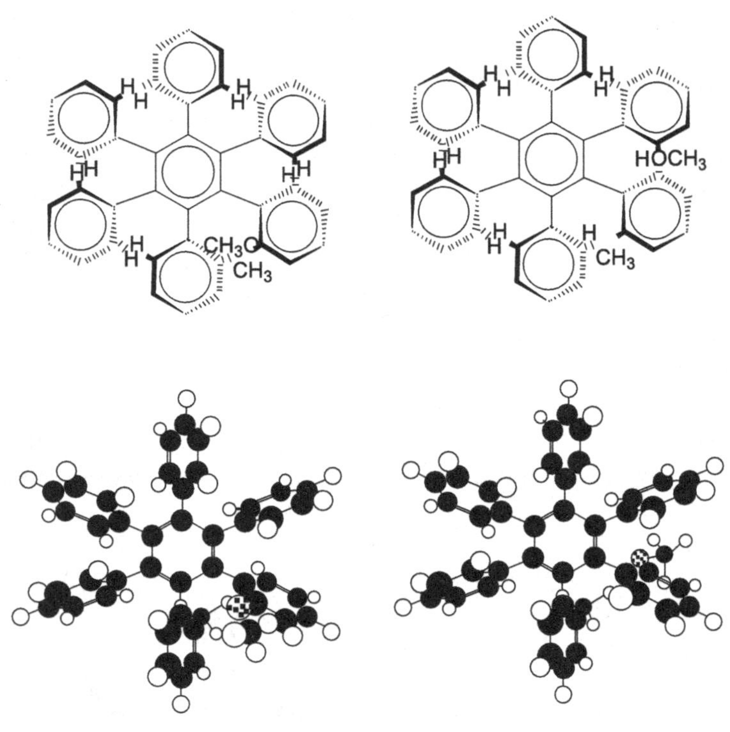

V
ortho-Disubstituted hexaphenylbenzene
exists as two diastereomers

Using tetraphenylcyclopentadienone (the diene) and diphenylacetylene (the dienophile) you will do a Diels-Alder cycloaddition reaction which is a bit different, as the initial adduct loses carbon monoxide to become aromatic. The excess dienophile will serve to ensure that all the diene is used in the reaction.

Tetraphenylcyclopentadienone   *Diphenylacetylene*   *Hexaphenylbenzene*

Warning! This experiment uses an open flame!!! Be sure that all flammable solvents have been removed from your laboratory work space!

PROCEDURE

1. Transfer 0.100 g of tetraphenylcyclopentadienone and 0.100 g of diphenylacetylene into a dry 13 x 100 mm test tube. Add 1 mL of high-boiling silicone oil.

2. Read the following NOTE, and then slowly heat the mixture to a gentle boil using a handheld microburner. Continue to gently boil the mixture for a total of fifteen minutes.

    [NOTE: Keep the flame moving around the test tube; boiling should occur in three to six minutes. After melting and dissolving into the hot silicone oil, the solution will be an intense, red-purple color. The color will fade, and the product will separate as tan crystals.]

3. After fifteen minutes of gentle boiling, cool the mixture to room temperature; then add 4 mL of hexane, and stir the mixture with a stirring rod.

4. Collect the crude product by vacuum filtration using a Hirsch funnel. Wash the solid with an additional 2 mL of hexane, and then wash with cold toluene (2 x 2 mL) to afford the white crystalline hexaphenylbenzene.

5. After air-drying in an open container in your drawer until the next laboratory period, weigh the product, and calculate your percent yield. <u>Do not take a melting point!</u> (It melts at >450°C, well above the maximum temperature of the Mel-Temp thermometers.)

INFRARED SPECTROSCOPY—PRELAB

Prelab Report: Due at the Beginning of the Lab Period

Name ____

Lab Section (Circle One): Mon Tues Wed Thur Fri AM/PM

For each pair of compounds predict the IR bands which are expected to be useful in distinguishing one compound from the other.

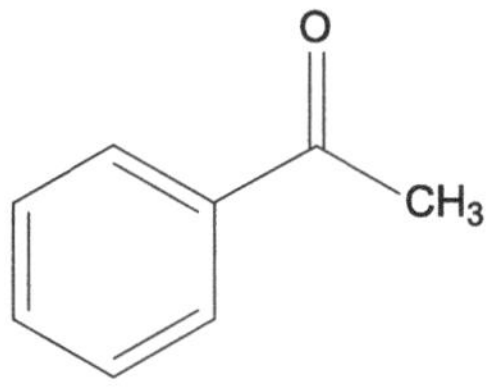

and

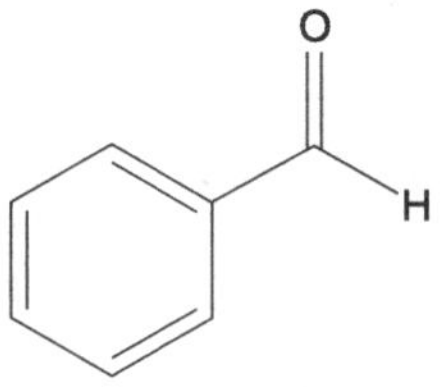

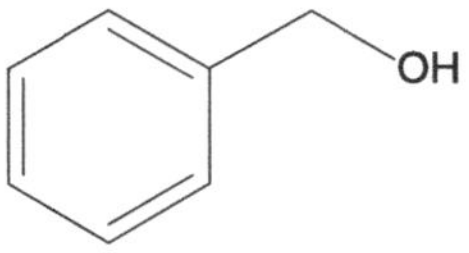

and

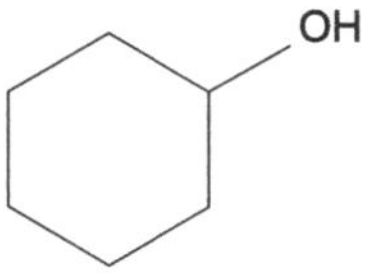

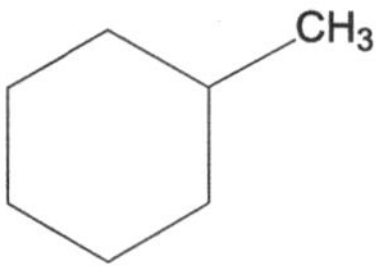

and

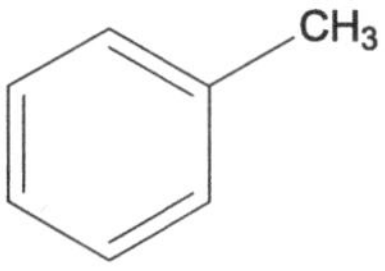

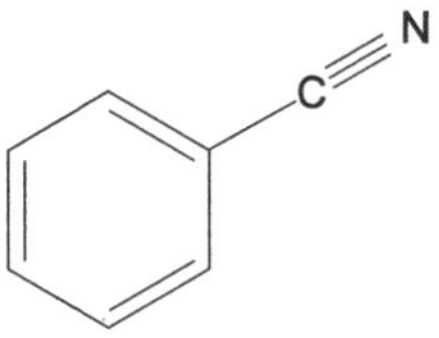

and

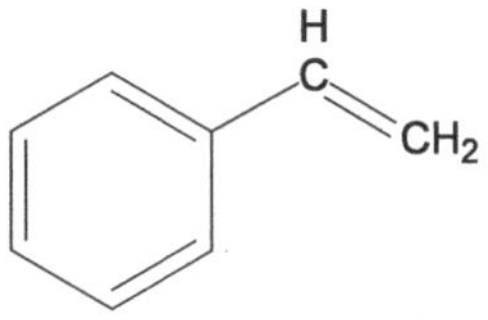

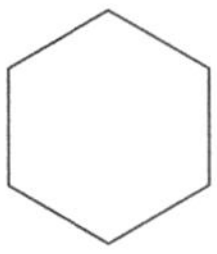

and

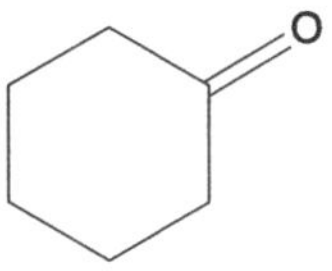

INFRARED SPECTROSCOPY

BACKGROUND

Infrared spectroscopy is one of the most useful methods for structure determination available to the organic chemist. The presence or absence of many common functional groups can be established unequivocally; the method does not destroy the sample; and only very small amounts of the sample are necessary. The utility is based on the fact that infrared radiation occurs at the same frequency as molecular vibration frequencies (5000 to 100 cm^{-1}). Identification of functional groups using infrared spectroscopy depends upon the empirical observation that many functional groups absorb at unique frequencies, regardless of the carbon skeleton to which they are attached. Besides the fundamental vibration frequencies, overtones (multiples of fundamental frequencies) and combination bands (additions of frequencies of two bands) also occur. Thus, even a simple molecule may possess a complex infrared spectrum.

Study of a great many molecules has led to the arbitrary division of the common infrared spectrum into several regions:

- The hydrogen stretching region

- The triple-bond region

- The double-bond region

- The C-H scissoring region

- The "fingerprint" region

The first four regions are useful in the identification of functional groups (OH, C=O, etc.) but are not useful in the identification of the individual compound. For example, since all nitriles (CN) absorb at the same frequency ($\sim$2250 cm^{-1}), it is easy to determine whether the functional group is present, but it is difficult to say which nitrile one actually has.

The "fingerprint" region is so named because a large number of C-C vibrations and other vibrations occur there in a manner that is not easy to predict. So many absorptions occur there, in fact, that this region can be used to identify individual compounds in the same way fingerprints identify individual people.

The table on the next page gives the frequencies of several common functional groups in an inert solvent (i.e., CCl$_4$). It should be emphasized that tables of this sort are empirical in nature. Discrepancies do occur, but an infrared spectrum provides a great deal of structural information about a compound.

One theoretical point should be kept in mind in the interpretation of infrared spectra: a change in dipole moment during the vibration is necessary if the vibration is to generate an infrared absorption band, and the intensity of the absorption depends upon the magnitude of the dipole moment change. Symmetrical acetylenes do not show a triple-bond stretching frequency in the infrared spectrum, and nearly symmetrical acetylenes may have an absorption that is so weak that it is difficult to distinguish the peak from the "background noise." On the other hand, a carbonyl group has a large permanent dipole moment; the carbonyl absorption band is usually one of the most intense bands in the spectrum.

Table 1 Infrared Maxima (cm⁻¹) for Common Functional Groups	
3650–3500	OH stretch
3600–3100	H-bonded OH stretch
3400–3200	NH stretch
3300	Acetylenic CH stretch
3100–3000	Olefinic CH stretch
3100–3020	Aromatic CH stretch
3000–2800	Aliphatic CH stretch
2700	Aldehydic CH stretch
~3000	Carboxylic OH (broad, very intense, ill-defined band; often overruns the CH stretch)
2450	Atmospheric CO_2
2250	Nitriles
2100	Acetylenes (CC triple bonds)
1735[a]	Saturated ester carbonyl stretch
1715[a]	Saturated ketone carbonyl stretch
1710[a]	Saturated aldehyde carbonyl stretch
1700[a]	Saturated acid carbonyl stretch (broad, ill-defined band)
1650[b]	Nonconjugated olefin C=C stretch
1600	Benzene ring breathing
1500–1400	C–H scissoring
1375	$C–CH_3$ (a C-H scissoring band)
1250–1000	C–O–C (intense bands)
975	*trans* H–C=C–H (usually good if there is also a 1650 band)
925–900	$C=CH_2$ terminal methylene (if there is also a 1650 band)
850–725	C-Cl stretch in carbon tetrachloride and chloroform
700	Monosubstituted phenyl ring

[a]If the carbonyl group is conjugated, subtract 20 cm⁻¹; if the carbonyl group is doubly conjugated, subtract 40–50 cm⁻¹.

[b]If the C=C is conjugated to another C=C or carbonyl, subtract 20 cm⁻¹.

SPECIAL INSTRUCTIONS

When you analyze an infrared spectrum, it is easier if you follow a series of steps:

1. Look first for the carbonyl (C=O) band. Look for a strong absorption at 1660–1820 cm⁻¹. This band is usually the most intense absorption band in the spectrum. If you see the carbonyl band, look for the other bands associated with functional groups that contain the carbonyl by going to step 2. If no C=O is present, go to step 3.

2. If a C=O is present, you will want to determine if it is part of an acid, an amide, an ester, an aldehyde, or a ketone.

   - ACID: Look for the presence of an O–H. It has a broad absorption near 3300–2500 cm^{-1}. This actually will overlap the C–H stretch.

   - AMIDE: The carbonyl absorption is less than 1700 cm^{-1}. Look for one (N–H) or two (NH$_2$) medium to weak bands, 3300–3500 cm^{-1}.

   - ESTER: Look for C–O single-bond absorption of medium intensity near 1000–1300 cm^{-1} (often two bands). There will be no O–H band.

   - ALDEHYDE: Look for the weak aldehyde-type C-H absorption bands at 2700–2850 cm^{-1} (often two bands).

   - KETONE: The carbonyl band will be at 1705–1725 cm^{-1} and nothing else.

3. If no carbonyl band appears in the spectrum, look for an alcohol O-H absorption.

   - ALCOHOL: Look for the broad, rounded OH band at 3300–3600 cm^{-1} and a C–O absorption at 1000–1300 cm^{-1}.

4. If no carbonyl or O-H bands are in the spectrum, check for the N-H of an amine.

   - AMINE: Look for one (N–H) or two (NH$_2$) medium to weak absorption bands in the 3300–3500 cm^{-1} region. NOTE: 3° amines have no N–H bands, as there are no Hs on the N!

5. If everything is negative so far, check for triple bonds.

   - NITRILE: Look for a weak CN peak about 2250 cm^{-1}.

   - ALKYNE: Look for a weak CC peak about 2100 cm^{-1}. If present check for the strong C–H stretch at 3300 cm^{-1}.

6. If everything is still negative, check for C=C double bonds from an aromatic or alkene.

   - ALKENE: Look for the weak C=C absorption near 1650 cm^{-1}. There also will probably be a C–H stretch band near 3050 cm^{-1}.

   - AROMATIC: Look for the medium benzene C=C absorption bands at 1450–1650 cm^{-1}. The C–H stretch band is much weaker than in alkenes but still at 3050 cm^{-1}.

7. If none of the previous groups can be identified, you may have one of the following:

   - ETHER: Look for a medium to strong C–O single-bond absorption at 1050–1260 cm^{-1}.

   - ALKANE: Look for a simple spectrum with C–H stretch around 2800–3000 cm^{-1} and C–H bends in the 1350–1450 cm^{-1} region.

PROCEDURE

1. Obtain from the instructor an unknown. Be sure to record the number of your unknown in your laboratory notebook. The unknown will be one of the following ten compounds:

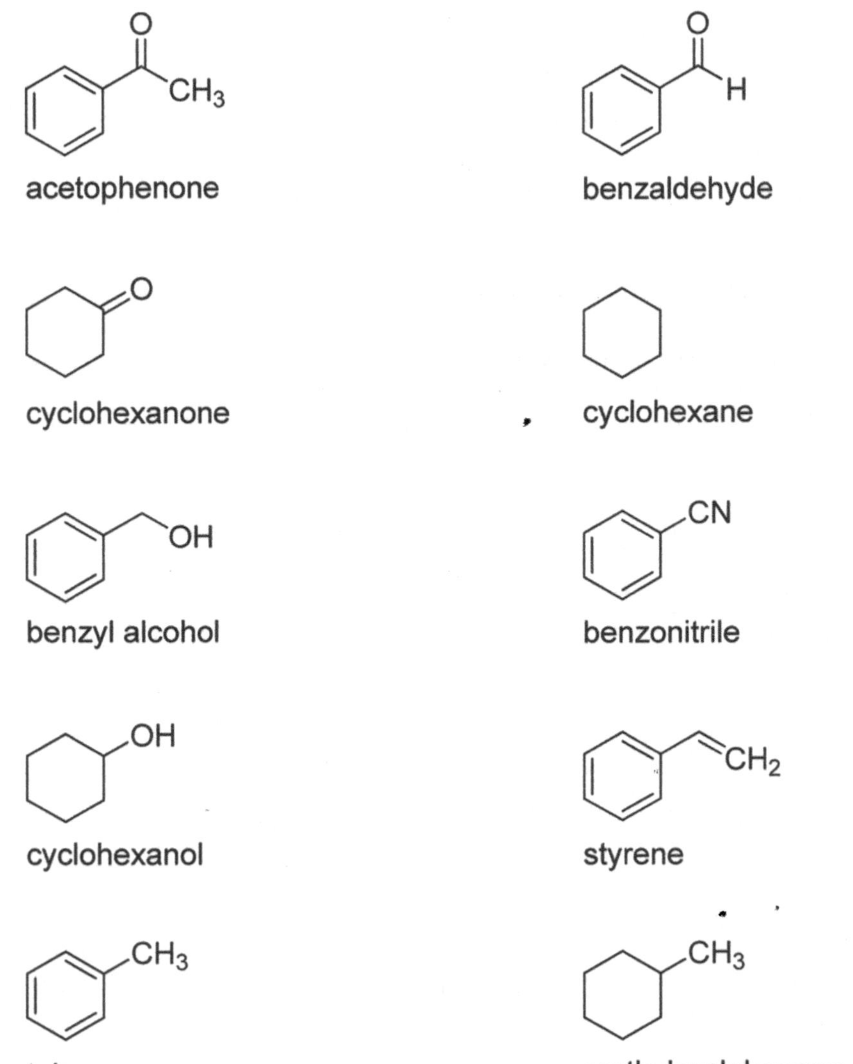

2. Obtain the infrared spectrum for your unknown, following the procedure explained by your laboratory instructor. (NOTE: Be sure to attach the infrared spectrum to your laboratory report.)

3. On the basis of the infrared spectrum for your compound, provide an argument in favor of its structure. Show which bands were used to verify the presence of particular structural features and how compounds with similar features were excluded from consideration.

Synthesis of Aspirin—Effect of the Catalyst—Prelab

Prelab Report: Due at the Beginning of the Lab Period

Name ____

Lab Section (Circle One): Mon   Tues   Wed   Thur   Fri   AM/PM

1. Classify each of the listed catalysts as a Lewis or Brønsted-Lowry acid or base.

   Sulfuric acid (H_2SO_4)

   Phosphoric acid (H_3PO_4)

   Magnesium bromide etherate ($MgBr_2 \cdot OEt_2$)

   Aluminum chloride ($AlCl_3$)

   Calcium carbonate ($CaCO_3$)

   Sodium acetate (NaOAc)

   Triethylamine (NEt_3)

   4-N, N-dimethylaminopyridine (DMAP)

2. How does a microwave cause something to heat up?

3. Physical properties:

   a. melting point of salicylic acid

   b. molecular weight of aspirin

   c. density of acetic anhydride

4. How much do two grains of pure aspirin weigh IN GRAMS?

5. If you started with 0.6753 g of salicylic acid and obtained an 83% yield of aspirin, how much does your final product weigh in grams? (Show your calculations.)

Microwave Aspirin Synthesis— Effect of the Catalyst

REACTION

In this experiment, aspirin (acetylsalicylic acid) will be prepared from salicylic acid and acetic anhydride, using either an acidic or a basic catalyst (or in some cases, no catalyst at all). Your catalyst will be assigned by your laboratory instructor and will be one of the following:

Sulfuric acid (H_2SO_4)	Calcium carbonate ($CaCO_3$)
Phosphoric acid (H_3PO_4)	Sodium acetate (NaOAc)
Magnesium bromide etherate ($MgBr_2 \bullet OEt_2$)	Triethylamine (NEt_3)
Aluminum chloride ($AlCl_3$)	4-N,N-dimethylaminopyridine (DMAP)

PROCEDURE

1. Place 0.69 g of salicylic acid and 1.5 mL of acetic anhydride in a microwave GlassChem vessel. Add your assigned catalyst. (If your catalyst is a liquid, add one drop; if it is a solid, add 0.02–0.04 g.) The initial reaction mixture generally forms a slurry; after microwave irradiation, it will become a homogenous solution. Subsequent formation of solids indicates the presence of the desired product.

2. Place the microwave tube into the turntable (be sure to record the number of the slot), and secure it. Your instructor will place the turntable into the MARS microwave reactor when all of the tubes are secured. The microwave reactor will then be set to 80°C and ten minutes of reaction time.

3. When the microwave reaction cycle is complete, remove your tube from the turntable, and stand it in a beaker for easy transport. As the tube cools, the acetylsalicylic acid should begin to crystallize from the reaction mixture. If it does not crystallize, scratch the walls of the microwave vessel with a glass rod to initiate precipitation. Once the tube has reached room temperature, continue cooling in an ice

bath for ten minutes. After crystal formation is complete, add 10 mL of cold water to the vessel and stir thoroughly. Collect the product by vacuum filtration, and wash twice with 2 mL of cold water.

3. Allow your sample to dry in your laboratory drawer at least overnight. Obtain a weight, melting point, and spectra (IR and ^{1}H NMR) of your product. Also analyze your product using the ferric chloride test for phenols as described below.

Ferric Chloride Test

This test looks for the presence of phenols and will show a positive result if there is any unreacted salicylic acid present in your product. Obtain three small test tubes. Add 0.5 mL of water to each test tube. Dissolve a small amount of salicylic acid (as much as will fit on the tip of your microspatula) in the first test tube and the same amount of your product in the second test tube. The third test tube just contains solvent and is used as the control. Add one drop of a 1% ferric chloride solution to each tube, shake well, and note the color. Formation of the iron-phenol complex gives a definite color ranging from red to violet.

POSSIBLE PROBLEMS

1. If your reaction mixture turned brownish, you probably added too much catalyst, which can cause polymer formation. It is advisable to start over.

2. If your reaction mixture turned viscous and yellowish, you may have added too much base (if your catalyst is a base), or you may just have added too much catalyst. It is advisable to start over.

REPORT

Your lab report should include the following:

1. Percent yield calculation and melting point for your product

2. IR and ^{1}H NMR spectrum of your product

3. Table of percent yields and purity of product for each catalyst (you will have to collaborate with your fellow students to gather this information)

4. Discussion of the relative merits of each catalyst and a recommendation for the best conditions for the synthetic transformation (defend your recommendation using the tenets of green chemistry)

REFERENCES

Montes, I.; D. Sanabria, M. Garcia, J. Castro, and J. Fajardo. 2006. *J. Chem. Educ* 83(4): 628–631.

Mirafzal, G. A., and J. M. Summer. 2000. *J. Chem. Educ.* 77(3): 356–357.

A BRIEF HISTORY OF ORGANIC COMPOUNDS RELATED TO ASPIRIN*

The antipyretic (fever reducing) property of the bark of the Willow tree (*Salix Alba*) was known to the ancient Greeks.

The underlying chemistry of the action of this natural material was unraveled in the 19th century. An examination of the developments surrounding organic compounds related to aspirin provides an interesting perspective on the state of development of this aspect of chemistry.

1763

Edward Stone noticed that chewing the bark of the willow tree helped to relieve the symptoms of malaria—chills and fever.

1827

The active ingredient in willow bark, *silicon*, was isolated.

Salicin

1831

A Swiss pharmacist, Johann Pagenstecher, distilled meadowsweet flowers and obtained and characterized a substance called **salicylaldehyde**. One variety of meadowsweet has the scientific name *Spiraea salicifolia*.

Salicylaldehye

1835

The German chemist, Karl Löwing, isolated **salicylic acid** (which he named Spärser) from a mixture of products obtained from the alkaline hydrolysis of **salicylaldehyde**.

Salicylic acid

1838

Raffaele Pivia, an Italian chemist, hydrolyzes **salicin** to produce glucose and **salicyl alcohol.** He further oxidizes salicyl alcohol to salicylic alcohol to salicylic acid, establishing a connection between that substance and the active ingredient in willow bark.

Salicyl alcohol

** This history draws heavily from the excellent book "Organic Molecules in Action" by Goodman and Morehouse (Gordon and Breach, 1973).*

1843

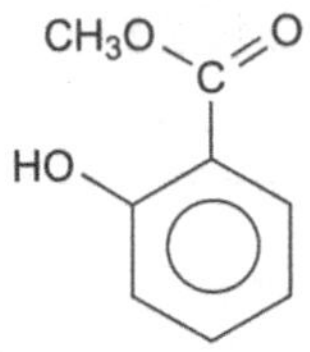

A related compound, **methyl salycilic**, was found by the French chemist, Auguste Cahours, and the American chemist, William Proctor, to be a major constituent of oil of wintergreen, which was extracted from the leaves of the wintergreen plant. Methyl salycilic continues to be used as a flavorant and a medicinal to this day. It is used in many medicines to relieve aching muscles.

Methyl Salicylate

1853

Charles Gerhardt of Strasbourg replaced the OH of salicylic acid with an acetyl group using acetic anhydride, the first synthesis of **acetylsalicylic acid**, which was later to become called aspirin.

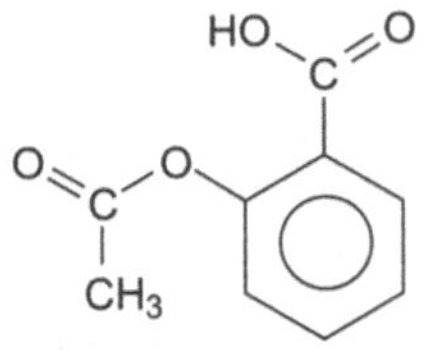

Acetylsalicylic acid
(Aspirin)

1859

Hermann Kolbe developed a convenient and inexpensive synthesis of salicylic acid.

1859–1893

During this period, salicylic acid which is moderately strong acid (pK_a = 3) was widely used as a medicine. The acid burned the mouth. Efforts to moderate the effects of its acidity resulted in the administration of the sodium salt of salicylic acid, sodium salicylate. The salt, however has an unpleasant taste.

1893

In an effort to find a less unpleasant way to administer salicylic acid, Felix Hoffman, a chemist working for the Bayer pharmaceutical company in Germany, reinvestigated the acetylating reaction first conducted by Gerhardt in 1853. Hoffman's father was rheumatic, which added a personal motivation for finding such a substitute.

The synthetic material called **aspirin** (a for acetyl, and the *spir* root, undoubtedly borrowed from the Latin name for meadowsweet) was shown to have all of the desirable properties of salicylic acid, but lacked the strong acidity of the acid and the unpleasant taste of the sodium salt.

ESSAY: ASPIRIN

Aspirin is one of the most popular cure-alls of modern life. Even though its curious history began over 200 years ago, we still have much to learn about this enigmatic remedy. No one yet knows exactly how or why it works, yet more than 15 billion aspirin tablets are consumed each year in the United States.

The history of aspirin began on June 2, 1763, when Edward Stone, a clergyman, read a paper to the Royal Society of London entitled "An Account of the Success of the Bark of the Willow in the Cure of Agues." By *ague*, Stone was referring to what we now call malaria, but his use of the word *cure* was optimistic; what his extract of willow bark actually did was to reduce the feverish symptoms of the disease. Almost a century later, a Scottish physician was to find that extracts of willow bark would also alleviate the symptoms of acute rheumatism. This extract was ultimately found to be a powerful **Analgesic** (pain reliever), **Antipyretic** (fever reducer), and **anti-inflammatory** (reduces swelling) drug.

Soon thereafter, organic chemists working with willow bark extract and flowers of the meadowsweet plant (which gave a similar compound) isolated and identified the active ingredient as salicylic acid (from *Salix*, the Latin name for the willow tree). The substance could then be chemically produced in large quantities for medical use. It soon became apparent that using salicylic acid as a remedy was severely limited by its acidic properties. The substance irritated the mucous membranes lining the mouth, gullet, and stomach. The first attempts at circumventing this problem by using the less acidic sodium salt (sodium salycilic) were only partially successful. This substance was less irritating but had such an objectionable sweetish taste that most people could not he induced to take it. The breakthrough came at the turn of the century (1893) when Felix Hofmann, a chemist for the German firm of Bayer, devised a practical route for synthesizing acetylsalicylic acid, which was found to have all the same medicinal properties without the highly objectionable taste or the high degree of mucosal membrane irritation. Bayer called its new product "aspirin," a name defined from a^- for acetyl, and the root *-spir*, from the Latin name for the meadowsweet plant, *spirea*.

Salicylic acid — Sodium salicylate — Acetylsalicylic acid (aspirin)

The history of aspirin is typical of many of the medicinal substances in current use. Many began as crude plant extracts or folk remedies whose active ingredients were isolated and structure were determined by chemists, who then improved on the original.

In the last few years, the mode of action of aspirin has just begun to be explained. A whole new class of compounds, called **prostaglandins**, has been found to be involved in the body's immune responses. Their synthesis is provoked by interference with the body's normal functioning by foreign substances or unaccustomed stimuli.

Prostaglandin E_2 — Prostaglandin $F_{2\alpha}$

These substances are involved in a wide variety of physiological processes and are thought to be responsible for evoking pain, fever, and local inflammation. Aspirin has recently been shown to prevent bodily synthesis of prostaglandins and thus to alleviate the symptomatic portion (fever, pain, inflammation, menstrual cramps) of the body's immune responses (that is, the ones that let you know something is wrong). One report suggests that aspirin may inactivate one of the enzymes responsible for the synthesis of prostaglandins. The natural precursor for prostaglandin synthesis is **arachidonic acid**. This substance is converted to a peroxide intermediate by an enzyme called **cyclo-oxygenase**, or prostaglandin syntheses. This intermediate is converted further to prostaglandin. The apparent role of aspirin is to attach an acetyl group to the active site of cylco-oxygenase, thus rendering it unable to convert arachidonic acid to the peroxide intermediate. In this way, prostaglandin synthesis is blocked.

Aspirin tablets (5-grain size) are usually compounded of about 0.32g of acetylsalicylic acid pressed together with a small amount of starch, which binds the ingredients. Buffered aspirin usually contains a basic buffering agent to reduce the acidic irritation of mucous membranes in the stomach, because the acetylated product is not totally free of this irritating effect. Buffering contains 0.325g of aspirin together with calcium carbonate, magnesium oxide, and magnesium carbonate as buffering agents. Combination pain relievers usually contain aspirin, acetaminophen, and caffeine. Extra-Strength Excedrin, for instance, contains 0.250g aspirin, 0.250g acetaminophen, and 0.065 g caffeine.

REFERENCES

"Aspirin Cuts Deaths after Heart Attacks," *New Scientist*, 118 (April 7, 1988): 22.

Collier, H.O.J. "Aspirin" *Scientific American*, 209 (November 1963):96.

Collier, H.O.J. "Prostaglandins and Aspirin," *Nature*, 232 (July 2, 1971):17.

Disla, E., Rhim, H.R.Reddy. A., and Taranta, A. "Aspirin on Trial as HIV Treatment," *Nature*, 366(November 18, 1993): 198.

Kingman's. "Will an Aspirin a Day Keep the Doctor Away" *New Scientist*, 117 (February 11, 1988):26.

Kolata, G."Study of Reye's-Aspirin Link Raises Concerns." *Science*, 227 (January 25, 1985):391

Macilwain, C.Aspirin on Trial as HIV Treatment." *Nature*, 364(July 29, 1993):369.

Nelson, N.A., Kelly, R. C., and Johnson, R. A. "Prostaglandins and the Arachidonic Acid Cascade." *Chemical and Engineering News* (August 16, 1982):30.

Pike, J.E. "Prostaglandins." *Scientific American*, 225 (November 1971):84.

Roth, G. J., Stanford, N., and Majors, P.W. "Acetylating of Prostaglandin Syntheses by Aspirin." *Proceedings of the National Academy of Science of the U.S.A.*, 72(1975): 3073.

Street, K. W. "Method Development for Analysis of Aspirin Tablets." *Journal of Chemical Education*, 65 (October 1988): 914.

Vane, J. R. "Inhibition of Prostaglandin Synthesis as a Mechanism of Action for Aspirin-like Drugs." *Nature-New Biology*, 231(June 23, 1971): 232.

Weiss Mann, g. "Aspirin." *Scientific American*, 264 (January 1991): 84.

Aspirin Synthesis—Alternate (Non-Microwave) Procedure

Prepare a hot water bath using about 50 mL of water in a 100-150-mL beaker. Adjust the water temperature to about 50°C. Weigh out 0.210 grams of salicylic acid and place in a dry 5-mL conical vial. Add 0.480 mL of acetic anhydride, followed by exactly one drop of concentrated phosphoric acid from a Pasteur pipet.

CAUTION: concentrated phosphoric acid is highly corrosive. Handle with care!

Add a magnetic spin vane and attach an air condenser to the vial. Clamp the assembly so that the vial is partially submerged in the hot water bath. Stir until the salicylic acid is dissolved, and then continue heating and stirring for another 10 minutes to complete the reaction.

Remove the vial from the hot water bath and allow it to cool. After it has cooled enough for you to handle, detach the air condenser and remove the spin vane with clean forceps. Place the conical vial in a small beaker (so it won't tip over) and allow to cool to room temperature. During the cooling period, the acetylsalicylic acid should begin to crystallize from the solution. If it does not, scratch the walls of the vial with a glass rod and cool the mixture slightly in an ice bath until crystallization has begun. At this point, if the vial is not already in an ice bath, put it in one and leave it there for ten minutes to complete the crystallization process.

Add 3.0 mL of water (measured in a 10-mL graduated cylinder) to the vial and stir thoroughly. Collect the solid by vacuum filtration, using an extra 1 mL of ice-cold water to assist in getting the last few crystals from the vial. When all the crystals have been collected in the Hirsch funnel, rinse them with several 0.5 mL portions of ice-cold water. Continue drawing air through the funnel for 5-10 minutes or until the crystals are nearly dry. Allow them to dry in your lab drawer until the next lab period.

Perform the ferric chloride test as instructed in the microwave procedure. On a TLC plate, please spot your starting material and your final product (both should be soluble in acetone or ethyl acetate) and develop this TLC plate in a chamber of 1:1 ethyl acetate–hexane.) Please calculate R_f, and sketch a copy of this this TLC plate into your lab report. Obtain the weight and melting point of your product.

QUESTIONS

1. What is the purpose of the concentrated phosphoric acid?

2. If you used 250 mg of salicylic acid and excess acetic anhydride in the preceding synthesis of aspirin, what would be the theoretical yield of acetylsalicylic acid in moles? In milligrams?

3. Most aspirin tablets contain five grains of acetylsalicylic acid. How many milligrams is this?

A GREENER SUZUKI REACTION—PRELAB

Name ____

Lab Section (Circle One): Mon   Tues   Wed   Thurs   Fri   AM/PM

1. What is the limiting reagent in this reaction?

2. Define "transmetallation" in terms of this reaction.

3. Provide the requested physical properties:

    a. melting point, 4-iodophenol

    b. molecular weight, phenylboronic acid

    c. density, methanol

    d. molecular weight, potassium carbonate

A GREENER SUZUKI REACTION[1]

BACKGROUND

There are relatively few synthetically useful carbon-carbon bond forming reactions. Commonly known ones include the Grignard reaction, the aldol condensation, and the Wittig reaction. Organometallic reactions can also be used for carbon-carbon bond formation; one such reaction is the Suzuki reaction, a popular and relatively mild method.

$$Ph\text{-}X + R'\text{-}BY_2 \xrightarrow[\text{base}]{Pd^0} Ph\text{-}R' + XBY_2 \text{ (where Y=OH)}$$

The mechanism involves the oxidative addition of an aromatic halide to the initial Pd^0 complex to form a Pd^{II} intermediate. The substituted boronic acid is activated with base, which converts the borane into the more reactive boronate, $Ph\text{-}B(OH)_3^-$. The slow step then is a transmetallation (so called because the nucleophile, R', is transferred from the activated boronic acid to the palladium, while the halide, X, becomes associated with the boron.) The newly formed Pd^{II} complex (with *both* organic attachments) undergoes a reductive elimination to form the coupled organic product. At the same time, the Pd^0 catalyst is regenerated and can be used again to continue the reaction cycle. See the illustration below, shown with hydroxide ion as the base.

Ph-R'

L-Pd-L

Ph-X

L　Ph
Pd
L　R'

transmetallation

L　Ph
Pd
L　X

(where L = H_2O ligands)

$XB(OH)_3^-$　　$R'\text{-} (OH)_3^-$

$R'\text{-}B(OH)_2 + OH^-$

---

[1]Modified from E. Aktoudianakis et al., "Greening Up" the Suzuki Reaction, *J. Chem. Educ.* **2008**, *85*(4), 555.

Suzuki reactions are often found as important steps in the synthesis of pharmaceutically relevant compounds. The original Suzuki reaction utilized refluxing benzene as the solvent;[2] currently these reactions are often performed using a mixture of organic solvent (such as THF) and water. However, one of the goals of green chemistry is to minimize the use of reagents that are hazardous to human health and the environment. One way of doing this is to use just water as the reaction solvent. Most organometallic reactions, such as the Grignard reaction and the Gilman reaction, cannot be done in the presence of water, but the Suzuki reaction works quite well even in an aqueous environment.

The reaction that you will be performing produces a biaryl product, in this case 4-phenylphenol, that is structurally similar to a number of non-steroidal anti-inflammatory products such as Diflunisal (used for arthritis treatment) and Felbinac (used as a topical treatment for muscle inflammation and arthritis).

PROCEDURE

CAUTION: Use caution when handling the materials used in this experiment. Wear gloves and appropriate eye protection at all times during this procedure. Phenylboronic acid and 4-iodophenol are skin irritants and harmful if inhaled. 10% Pd/C is harmful if swallowed. Methanol is flammable and toxic if swallowed. HCl causes burns and is irritating to the respiratory system.

Place 0.122 g phenylboronic acid, 0.415 g solid potassium carbonate, 0.220 g 4-iodophenol, and 10 mL of water into a 50-mL roundbottom flask with a stir bar. Add 1 mL of water to the supplied vial containing 3 mg of 10% Pd/C. Swirl to create a suspension, and add to the roundbottom flask.

Attach a water-cooled condenser and reflux the solution vigorously for 30 minutes, maintaining rapid stirring. Some solid may precipitate during the course of the reaction. When 30 minutes at reflux have elapsed, remove the roundbottom from the heat and allow to cool to room temperature.

When the solution is at room temperature, add 2M HCl until the solution is acidic (blue litmus turns pink). Collect the crude solid, which still contains the catalyst, by vacuum filtration. Wash the collected solid with 10 mL of water.

---

[2]Miyaura, N.; Yanagi, T., Suzuki, A. *Synth. Commun.* **1981**, *11*, 513.

Dissolve the semi-purified solid (not everything will dissolve) in 10 mL of methanol in a 25-mL Erlenmeyer flask and remove the Pd/C catalyst by gravity filtration (fluted filter paper) into a 50-mL Erlenmeyer flask. Add 10 mL of distilled water to the methanol solution, causing the solid product to precipitate. Heat until the entire product has redissolved, adding 1–2 mL portions of methanol if necessary. (You are performing a mixed-solvent recrystallization.) Once the solid is completely dissolved, allow the solution to cool slowly to room temperature, followed by an additional 10 minutes in an ice bath to complete crystallization.

Collect the recrystallized product by vacuum filtration and allow to dry (store in your drawer) until the following laboratory period. Obtain a weight, melting point, and IR spectrum.

QUESTIONS

1. Discuss the benefits of performing the Suzuki reaction under aqueous conditions. Compare and contrast this reaction with the method utilized by Callam and Lowary (*J. Chem. Educ.*, 2001, 78 (7), p. 947)—article available at http://pubs.acs.org/doi/pdf/10.1021/ed078p947 when on campus.

2. Why is it important to develop reactions that use water as the solvent?

3. Compare the Suzuki coupling reaction to the Grignard reaction. Points to consider include: amount of metal used (catalytic or stoichiometric); functional groups tolerated in the substrates; and solvent considerations.

4. Calculate the % atom economy for the reaction (note that only reactants that are consumed should be included in the calculation):

   % atom economy = (molecular weight of product/sum of molecular weight of reactants) x 100

5. Does the % atom economy improve if we used 4-bromophenol instead of 4-iodophenol as a reactant?

LAB REPORT

Include the weight, melting point, and percent yield calculation. Attach the IR of your product with a basic analysis. Answer the questions above.

IODINATION OF SALICYLAMIDE—PRELAB

Name ___________________________________

Lab Section (Circle One): Mon   Tues   Wed   Thurs   Fri   AM/PM

1. What is the limiting reagent for this reaction?

2. Give a mechanism for the reaction, assuming that I^+ is the electrophile.

3. What color do you expect the solution to be at the end of the reaction?

4. Provide the requested physical properties:

    a. melting point, salicylamide

    b. molecular weight, salicylamide

    c. density, absolute ethanol

    d. boiling point, absolute ethanol

IODINATION OF SALICYLAMIDE[1]

BACKGROUND

The iodination of benzene is an example of *electrophilic aromatic substitution* (a proton on the aromatic ring is replaced by an iodine). The electrophile [I$^+$] in this reaction is formed by the reaction of sodium hypochlorite (NaOCl, bleach) with sodium iodide (NaI). The iodine cation can add to the benzene ring, <u>temporarily</u> suspending the aromatic resonance (although the intermediate arenium ion is still somewhat stabilized by resonance). Because aromatic stabilization of the benzene has been lost, this step (addition of the electrophile) is going to be the rate-determining step. Aromaticity is then restored by removal of a proton from the sp^3 ring carbon, so the proton loss will be fast. The sequence is shown below:

MECHANISM FOR IODINATION OF BENZENE

$$NaOCl + NaI \rightarrow I^+$$

Step 1

Sodium hypochlorite reacts with sodium iodide to produce iodine cations.

Step 2

The iodine cation reacts with the π cloud of the arene and adds to the ring. (This is the rate-determining step; the aromaticity of the ring is broken.)

Step 3

Water acts as a Lewis base to restore the aromaticity of the ring (the proton from the sp^3 carbon is removed).

---

[1]   Modified from Eby, E. and Deal, S.T. "A Green, Guided-inquiry Based Electrophilic Aromatic Substitution for the Organic Laboratory", *Journal of Chemical Education*, 85(10) 2008.

Substituted benzene compounds react via the same mechanism. In principle one might expect that any (or all) of the hydrogens could be replaced; however the substituent group has two effects on the reaction:

1) The *Rate of Reaction* (compared with benzene) will be different. If the reaction is slower, the substituent is said to <u>deactivate</u> the ring (towards electrophilic substitution); if the reaction is faster, the substituent is said to <u>activate</u> the ring.

2) The *Orientation of Reaction* is indicated by where the electrophile actually ends up (relative to the substituent). That is, the substituent 'directs' the incoming electrophile preferentially to certain positions (relative to it). In general, substituent groups that can donate electron density into the ring lead to a faster reaction (compared to benzene) and the incoming electrophile goes to the 2- and 4- positions (relative to the substituent). Conversely, substituent groups that withdraw electron density out of the ring lead to a slower reaction (compared to benzene) and the incoming electrophile goes to the 3- position (relative to the substituent). Your lecture text will delve deeper into the theory behind these observations.

In this experiment, you will perform an iodination of salicylamide. Upon completion of the reaction and isolation of the product, your task will be to determine which product was actually made (since there are potentially four different products). You will use infrared spectroscopy to make the determination.

Determining Substitution Patterns Using IR

Infrared spectroscopy is commonly used for determining the presence (and absence) of functional groups in organic compounds. While this is important information, there is a wealth of additional information that can be determined via IR as well. We often do not consider this information because it is generally to be found in the "fingerprint" region of the IR spectrum and can be difficult to identify unambiguously.

The substitution pattern of aromatic rings is something that can be identified using IR fairly easily. The table below will guide you in determining the identity of your product (which, of course, is one of the trisubstituted choices). There is a similar, somewhat more detailed, table in Appendix IV of your laboratory manual.

Ring Substitution Pattern	Expected Peaks (cm^{-1})
Monosubstituted	770-715 (strong)
1,2-Disubstituted	770-730 (strong)
1,3-Disubstituted	820-760 (strong)
1,4-Disubstituted	870-800 (strong)
1,2,3-Trisubstituted	790-750 (strong)
1,2,4-Trisubstituted	850-800 (strong)
1,3,5-Trisubstituted	910-830 (strong)

PROCEDURE

CAUTION: Use caution when handling the materials used in this experiment. Wear gloves and appropriate eye protection at all times during this procedure. Salicylamide and sodium iodide are irritants. Sodium hypochlorite and hydrochloric acid are irritants and are corrosive.

Measure out 0.250 g of salicylamide and place into a 25-mL round-bottom flask. Dissolve the salicylamide in 5 mL of absolute ethanol, warming the flask with your hand to speed up the dissolution.

Once the salicylamide is completely dissolved, add 0.3 g of sodium iodide (NaI) to the reaction mixture, stirring with a glass stirring rod until the solution is homogeneous. Place the 25-mL round-bottom containing the reaction mixture (with an air condenser attached) into an ice bath. (Think about thermal contact!! Ice bath means ice AND water.)

When the reaction is cooled to 0 °C (about 10 minutes), remove the reaction vessel from the ice bath and obtain 9.2 mL of 6% (w/v) sodium hypochlorite solution (ultra strength household bleach) from the hood (your TA will be dispensing it). Quickly add the bleach directly into the reaction mixture using a Pasteur pipet through the air condenser. Swirl the flask vigorously ONCE to completely mix the contents and then let it rest on the benchtop. The solution will change colors from the initial clear reaction mixture to a dark red-brown to increasingly lighter shades of yellow. When the solution reaches a faint, pale yellow color, the reaction is complete. (Typically, this takes less than five minutes.) Allow the reaction vessel to sit on the benchtop undisturbed for 10 minutes.

Add 2.5 mL of a previously prepared solution of 10% (w/v) sodium thiosulfate to the reaction solution and swirl the flask until the contents are thoroughly mixed. Next, CAREFULLY acidify the reaction solution by slowly adding 10% HCl dropwise, until the solution is just acidic to blue litmus paper (do not add too much acid!). You will notice a white solid beginning to form in the reaction vessel. At this point, the pH of the solution is near the desired acidity. Continue adding 10% HCl if necessary, but carefully monitor the acidity.

Once the mixture is acidic, filter using vacuum filtration and a Hirsch funnel. When the initial filtration is complete, turn off the vacuum, add 2 mL of water to the funnel, stir up the solid to dissolve any excess NaCl by-product, and then continue the filtration process. Collect the precipitate from the frit and recrystallize from 95% ethanol. (After the recrystallization mixture has cooled to room temperature, place it an ice bath for 25 minutes to complete crystallization.) Filter using vacuum filtration and allow the crystals to remain on the funnel with air being drawn over them for another 25 minutes to speed up the drying process.

Once your product is dry, weigh it and collect an IR spectrum. (Ask your TA for help if necessary.) Be sure to carefully label all peaks between 700 – 900 cm^{-1} on your spectrum, especially the largest peak in this region.

QUESTIONS

1. What are the possible iodination products?

2. Predict the most likely site of iodination and explain your reasoning.

3. What product was actually produced? How do you know this?

LAB REPORT

Include the weight, melting point, and percent yield calculation. Attach the IR of your product, and indicate which peaks allowed you to determine the product. Answer the questions above.

MICROWAVE SYNTHESIS OF A NATURAL INSECTICIDE—PRELAB

Prelab Report: Due at the Beginning of the Lab Period

Name ____

Lab Section (Circle One): Mon   Tues   Wed   Thur   Fri   AM/PM

1. What *is* Montmorillonite K-10? (Structure? Composition?)

2. The product is referred to as MDP Insecticide. What does the MDP stand for?

3. Discuss the "green chemistry" aspects of this experiment, with particular attention to atom economy and use of benign/recyclable components. Also, explain why the use of a microwave adds to the "greenness" of the experiment.

4. Provide the requested physical properties:

Molecular weight of sesamol

Boiling point of 3-methyl-2-butenal

Density of ethyl acetate

Microwave Synthesis of a Natural Insecticide

REACTION

sesamol + 3-methyl-2-butenal $\longrightarrow$ $\rightleftharpoons$ $- H_2O$

MDP

PROCEDURE

Preparation of the K-10 Catalyst

1. Weigh 0.5 g of Montmorillonite K-10 clay and place in a 50-mL Erlenmeyer flask. Add 10 mL of saturated aqueous K_2CO_3 to the flask. Add a magnetic stir bar, and stir at room temperature for thirty minutes. When stirring is complete, collect the clay by vacuum filtration. Allow the clay to dry in the funnel for several minutes; then transfer to a watch glass, spread out into a thin layer, and continue drying in a 110° C oven for one hour. When the drying period has elapsed, remove the clay from the oven and allow to cool to room temperature in a desiccator.

Synthesis of the Insecticide

1. While the clay is cooling, weigh 0.138 g of sesamol and place in a microwave GlassChem vessel. Add 106 µL of 3-methyl-2-butenal to the container. Finally, add all of the cool, dry clay, and mix thoroughly with a metal spatula. Cap the microwave vessel, place it into the turntable (be sure to record the number of the slot), and secure it.

2. Your instructor will place the turntable into the MARS microwave reactor when all of the tubes are secured. The microwave reactor will then be set to 80°C and eight minutes of reaction time.

3. When the microwave reaction cycle is complete, remove your tube from the turntable, and stand it in a beaker for easy transport. Allow the reaction mixture to cool to room temperature.

4. Add 5 mL of ethyl acetate to the microwave vessel and swirl gently. Separate the desired product (now dissolved in the ethyl acetate) from the clay catalyst by vacuum filtration. Rinse the clay twice with 5 mL of ethyl acetate to collect as much crude product as possible. Place the washed clay in the designated container for recycling.

5. Transfer the ethyl acetate to a <u>tared</u> 25-mL round-bottom flask with a magnetic stir bar, and remove the solvent by simple distillation. Place the recovered ethyl acetate in the designated container in the hood for reuse. Residual ethyl acetate may be removed by gently blowing a stream of air over the crude product until the mass is constant. Weigh the flask containing the crude product (it should be a dark brown, viscous oil), and calculate a percent yield. Analyze the product by TLC, IR, and/or ^{1}H NMR spectroscopy.

REPORT

Your laboratory report should include the following:

1. Detailed mechanism for the synthesis

2. Discussion of the "green" aspects of this procedure

REFERENCES

Dintzner, M. R ., P. R. Wucka, and T. W. Lyons. 2006. *J. Chem. Educ.* 83(2): 270–272.

MOLECULAR MODELING EXPERIMENT—THE EFFECT OF MOLECULAR STRUCTURE ON CONJUGATION

BACKGROUND

The term "conjugation" refers to the interaction of adjacent functional groups containing pi bonds and/or p-orbitals. Conjugation can affect both the chemical and physical properties of molecules. There are many ways to describe conjugation—resonance theory and orbital hybridization theory are the two most common methods. A classic example of conjugation/resonance theory is the Kekulé method of drawing benzene.

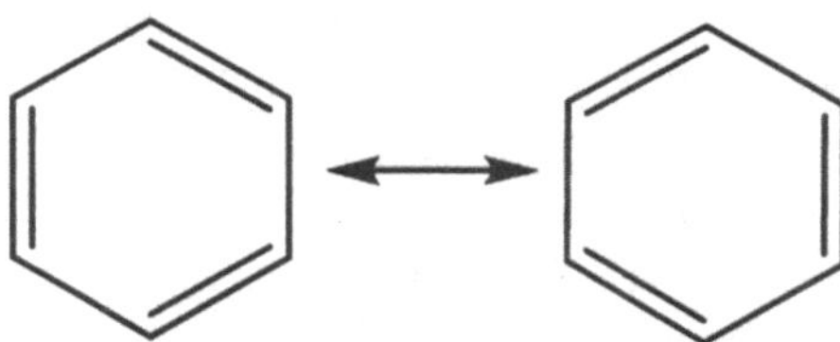

Figure 1 Kekulé Structures of Benzene

We can extend the concept of conjugation to look at compounds containing multiple rings and/or functional groups. Consider biphenyl as an example. By looking at the resonance structures (or at the molecular orbital picture) we can conclude that the two rings must be coplanar, as that is the only way to achieve the maximum orbital overlap and hence the greatest stabilization.

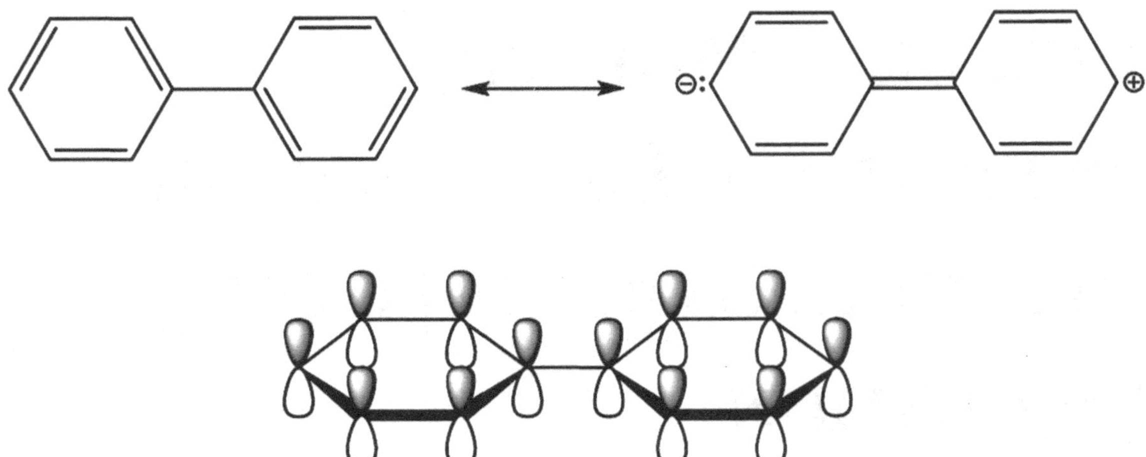

Figure 2 Biphenyl Structures

The effects of conjugation can be detected using many different types of measurement. One of the most common is UV-vis spectroscopy. UV-vis focuses on the section of the electromagnetic spectrum with wavelengths between 200 and 800 nm. The energy of light in this region is such that when it is absorbed by a compound, an electron is generally promoted from the HOMO (highest occupied molecular orbital) to the LUMO (lowest unoccupied molecular orbital), creating an excited state. The excited state is inherently unstable and quickly decays back to the ground state, emitting a photon in the process. The UV-vis spectrophotometer detects this emission and creates a spectrum showing the relationship between wavelength and intensity.

Electrons in organic molecules are generally described as residing in pi (π), sigma (σ), or nonbonding (n, often a lone pair) orbitals. The empty orbitals are commonly indicated as π^* and σ^*. Possible transitions resulting in absorption in the UV-vis spectrum are (in order of decreasing energy): σ to σ^*, n to σ^*, π to π^*, and n to π^*. In general, the presence of conjugation in a molecule will cause the wavelength of maximum absorption (λ_{max}) to be shifted to longer wavelengths (lower energy). Molecules with extensive conjugation also have larger molar extinction coefficients (ε; recall Beer's law: A= εbC).

In this exercise, we will investigate the effects of steric hindrance on conjugation as shown by UV-vis spectroscopy for several related organic compounds.

QUESTIONS

1. The UV-vis spectrum of benzene has a major band at λ_{max}=204 nm (ε=7400). The absorption band of mesitylene (1,3,5-trimethylbenzene) is somewhat shifted to a longer wavelength, λ_{max}=210 nm (ε=7000). The UV-vis spectrum of biphenyl is quite different, however, showing an intense peak at λ_{max}=246 nm (ε=16300). But the UV-vis spectrum of dimesityl (see structure below) shows no peak at all in this range, and in fact is quite similar to that of mesitylene. Why is that?

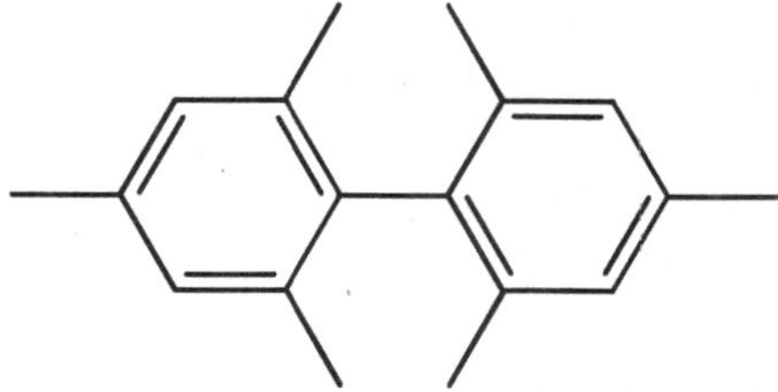

Figure 3 Dimesityl

2. Provide minimized structures (Chem3D) that clearly show the relationships in space between the nitro group and the benzene ring. Explain the observed absorption maxima.

   a. nitrobenzene, λ_{max}=252 nm (ε=8620)

   b. 2-nitrotoluene, λ_{max}=250 nm (ε=5950)

   c. 2-isopropylnitrobenzene, λ_{max}=247 nm (ε=3760)

   d. 2-*t*-butylnitrobenzene, no peak in this region

3. Provide minimized structures (Chem3D) of the following compounds that clearly show the relationships in space between the benzene rings and the double bond. Explain the observed absorption maxima.

   a. *Trans*-stilbene, λ_{max}=295 nm (ε=25000)

   b. *Cis*-stilbene, λ_{max}=283 nm (ε=12300)

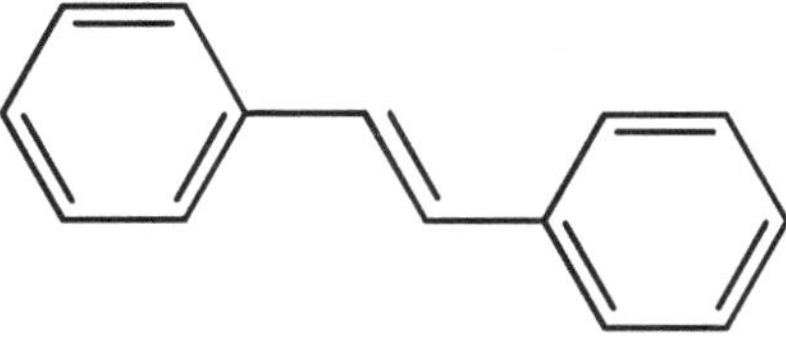

Figure 4a Trans-stilbene

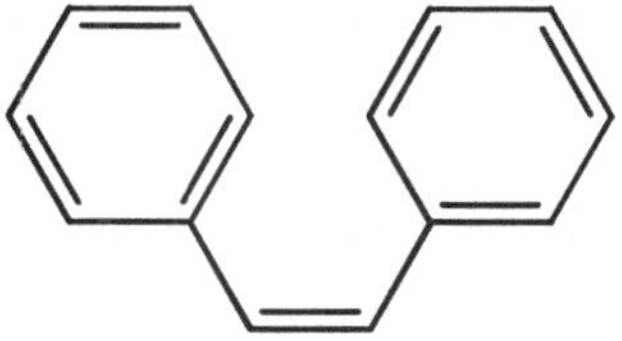

Figure 4b Cis-stilbene

How to Draw in the ChemDraw Window

Click on the Benzene Ring tool in the toolbar (it looks like a benzene ring, big surprise). Then click in the drawing space. You will get a benzene ring. To add an alkyl substituent, click on the Solid Bond tool (it looks like a straight line on the toolbar) and then put the cursor over one of the carbons on the benzene ring; you should see a "hit box" (on my computer it shows up as green). With the hit box active, click. You should sprout a bond off of the benzene ring, giving you toluene. To draw a nitro group, double-click at the end of the methyl group, and type "NO_2," into the resulting box. Then use the Select tool (it looks like a lasso) to select the NO2, click on the Structure menu, and choose Expand Label.

How to Minimize Your Drawing and Get Its Energy

Click on Calculations, MM2, Minimize Energy, Run. Your final energy will appear at the bottom of the screen.

REPORT

Your report should include the following:

1. Title page

2. Abstract

3. Answers to problems, including structures and written analyses

4. Conclusions

MOLECULAR MODELING EXPERIMENT—ROTATIONAL STUDIES OF 2-BROMOBUTANE AND INQUIRIES TOWARDS E2 ELIMINATIONS

BACKGROUND

One common method for the synthesis of alkenes is an E2 elimination of HX (where X can be Br, Cl, or I) from an alkyl halide in the presence of a strong base. The proton that is lost is referred to as the β proton, as it is attached to the carbon adjacent to that bearing the halide functional group. Many alkyl halides contain more than one type of β proton, leading to multiple products. The product distribution in the E2 elimination can generally be predicted using Zaitsev's rule, which states that in cases where more than one alkene can be formed in a reaction, the major product is generally the most stable one. This is illustrated below, in the E2 elimination of 2-bromobutane in the presence of potassium t-butoxide:

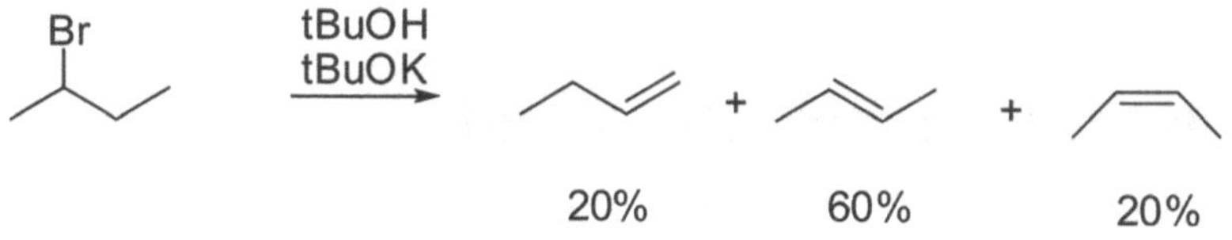

The E2 elimination requires that the β proton and the leaving group (in this example the Br) be either *anti-periplanar* (this means that the dihedral angle between proton and Br is 180°; see Newman projections 1 and 2 below) or *syn-periplanar* (the dihedral angle is 0°; see Newman projection 3 below). More details on this reaction can be found in the course textbook.

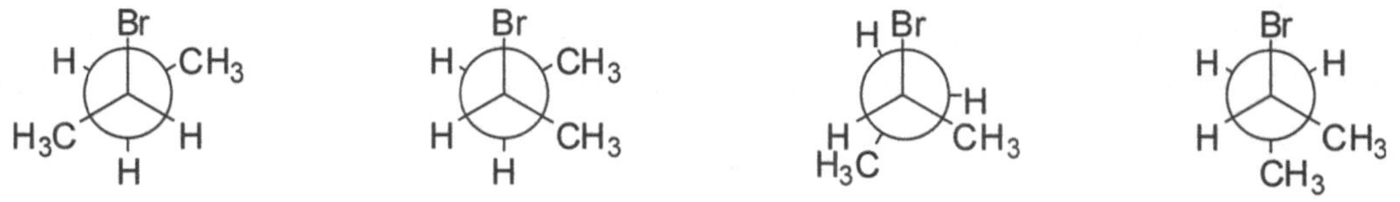

The purpose of this exercise is to investigate the observed predominance of the anti-periplanar elimination and of the *trans* alkene product using molecular-modeling techniques.

COMMENTS

The E2 reaction is considered to be controlled by kinetics rather than by thermodynamics, because the reaction is irreversible under standard conditions. Kinetic control means that the products that are formed fastest will be present in the greatest amount. Product stability (a thermodynamic property) will have no direct bearing on the product distribution. Instead, the differences in activation energies for the various reaction pathways would allow us to predict the product distribution. Basic molecular-modeling techniques do not allow us to estimate the energy of the transition states; however, we can estimate the relative amounts of the "starting materials" (the different conformations of the starting material). If there is little difference in activation energy of the various pathways, then the relative amounts of the various conformations will give us a good approximation of the product distribution. If, on the other hand, there is a large difference in activation energy of the various pathways, then the energy differences of the conformers may not accurately reflect the ratio of products.

INSTRUCTIONS

1. Draw the three alkenes (1-butene and *cis/trans* 2-butene) in the ChemDraw Window, and minimize using Chem3D. Record the minimized energies.

2. Draw 2-bromobutane in the ChemDraw Window, and minimize using Chem3D. Record the minimized energy.

3. Rotate your 2-bromobutane structure about the C2-C3 bond, 60° at a time, until a full 360° has passed. Draw all of the Newman projections (note that ChemDraw includes templates for both staggered and eclipsed Newmans), and record the corresponding energies.

How to Draw in the ChemDraw Window

Click on the Solid Bond tool in the toolbar (it looks like a straight line). Then click in the drawing space. You will get a single bond. To lengthen the chain, put the Solid Bond tool at the end of that single bond; you should see a "hit box" (on my computer it shows up as green). With the hit box active, click again. You should sprout another bond off of the first one, giving you three carbons total. Do that one more time, and you have butane.

To go from a single to a double bond, put the Solid Bond tool over the middle of the bond in question. You should see a rectangular hit box appear. Click once, and it will turn into a double bond.

Use this method to draw your three alkenes. To draw 2-bromobutane, use the same method to draw butane, and then sprout a branch off of carbon 2 using the Solid Bond tool. To convert the carbon to bromine, put the Solid Bond tool over the end of the bond (hit box active), and press the B key. This is the shortcut for putting in bromine. (There are other shortcuts: O, P, S, L are some of the common ones. Feel free to experiment.)

How to Minimize Your Drawing and Get Its Energy

Click on Calculations, MM2, Minimize Energy, Run. Your final energy will appear at the bottom of the screen.

How to Do the Rotations

Use the Select tool (the left-most tool, a black arrow) to select the center bond. Then rotate the molecule so that it looks like a Newman projection (see previous pictures). To get the energy of a particular conformation, use Calculations, MM2, Compute Properties, Run (not Minimize Energy).

Once you have the energy of your starting conformation, click on the Trackball tool (third from the left, a circle with an arrow). Right next to the Trackball is a downward-facing triangle (very small). Click on that, and you get something that looks kind of like a compass (see screenshot below).

Click on the line with a dot (second from the right)—this selects dihedral rotation, which means rotation about a selected bond. And you already selected the center bond. In the window below the compass, type in 60 (you want to do 60° rotations), and hit enter. The picture will change to the next Newman projection (which will be eclipsed). Calculate the energy as before, and then repeat the 60° rotation etc., until you have gone all the way around. At this point, you have collected all the data necessary to answer the questions.

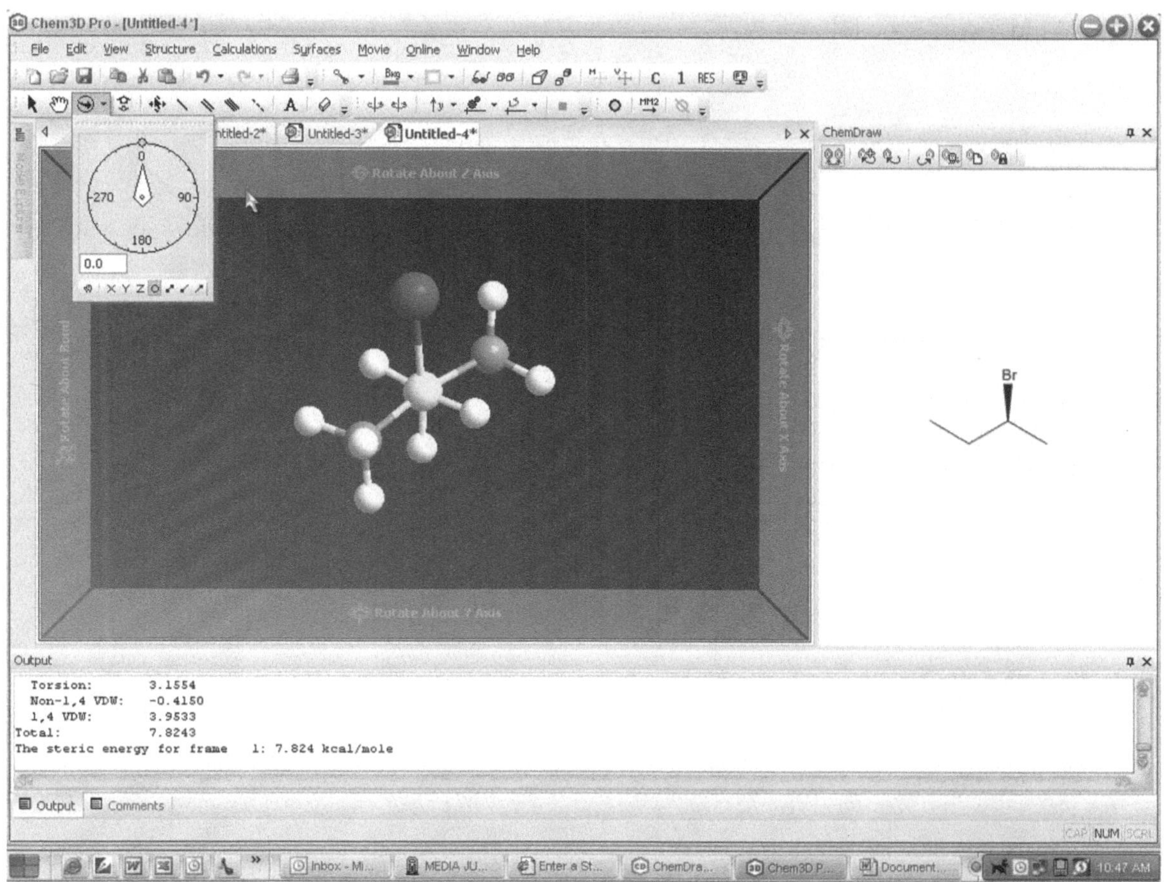

QUESTIONS

1. Look at the four Newman projections shown above. Draw the structure of the alkene that would be formed by E2 elimination from each conformation. Note that conformation 4 does not seem to show a hydrogen anti to the bromine, but keep in mind that there are other carbons in the molecule beyond the two that have been expanded in the Newman projection.

2. Calculate the ratio of the syn-periplanar conformation (3) to the anti-periplanar conformation leading to the *trans* alkene. (Is that conformation 1 or 2? See your answer to #1 above.) Does this ratio explain the predominance of the anti elimination in the E2 mechanism?

3. Calculate the ratio of the anti-periplanar conformations leading to *cis* and *trans* alkenes. Does this ratio explain the predominance of the *trans* alkene in the observed product distribution?

4. Calculate the difference in energy (ΔE) between the *cis* and *trans* alkenes produced in this reaction. Why shouldn't this number be used to explain the observed product distribution?

5. Use a spreadsheet program such as Excel to plot a rotational energy diagram for 2-bromobutane, using the energies calculated in Chem3D.

REPORT

Your report should include the following:

1. Title page

2. Abstract

3. Data (includes bond-line and Newman structures with corresponding energies—be sure to specify units)

4. Answers to problems, including calculations

5. Rotational energy diagram

6. Conclusions

Molecular Modeling Experiment—Friedel-Crafts Alkylation Reaction

BACKGROUND

The Friedel-Crafts reaction is one of a family of reactions classified as electrophilic aromatic substitution (EAS), whose mechanisms have been studied extensively. One aspect of the mechanisms that can be investigated using molecular modeling is the effect of *steric hindrance*. As an example, compare the methylation of *t*-butylbenzene with that of toluene. The reaction of *t*-butylbenzene with chloromethane and aluminum chloride gives primarily the *para*-methyl-*t*-butylbenzene product, with very little of the *ortho* product being observed. On the other hand, the same reaction with toluene gives approximately equal amounts of *ortho*- and *para*-dimethylbenzene. How do we explain this observation? We cannot use molecular mechanics to calculate the energies of the transition states, but it is generally believed that the energies of reactive intermediates (such as carbocations) are often good approximations of the energies of the corresponding transition states.

Representations of the transition states as well as the corresponding cation intermediates for *ortho* and *para* methylation of *t*-butylbenzene are shown below.

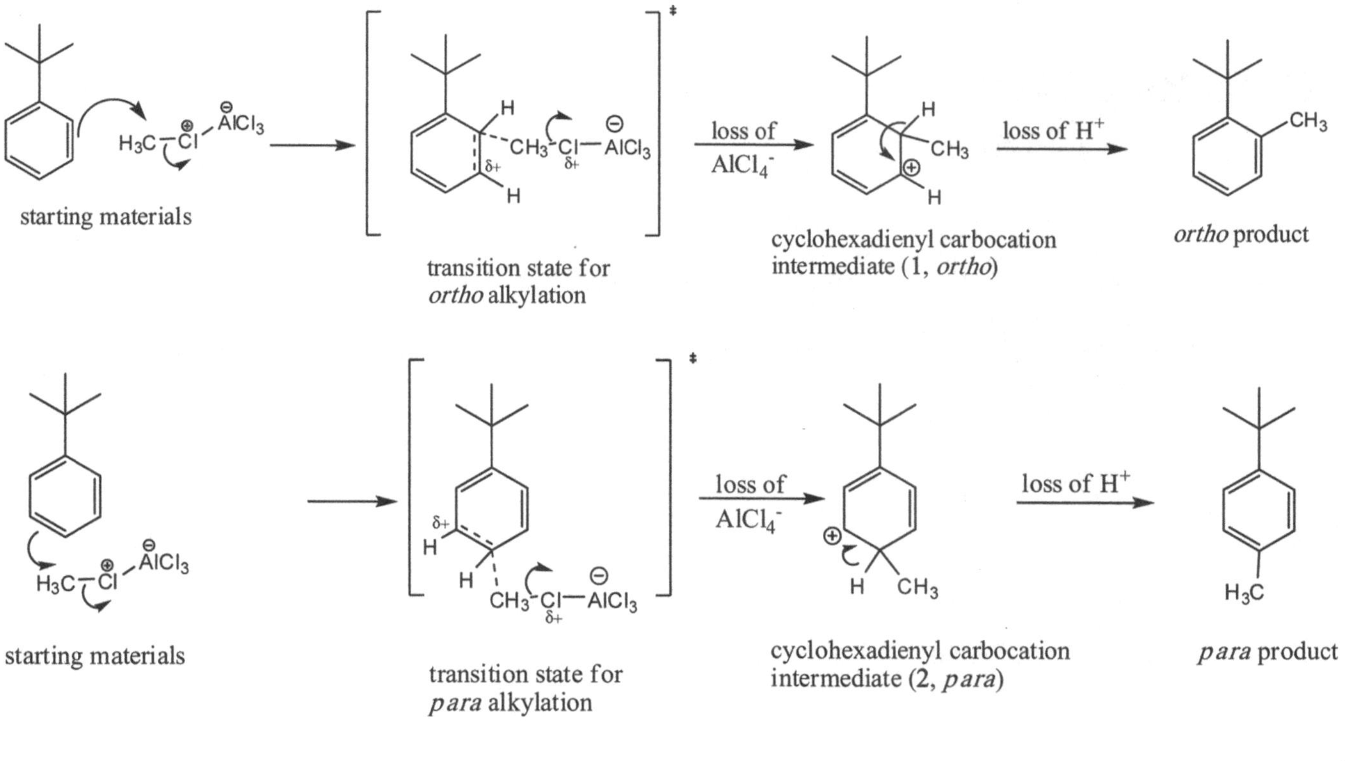

COMMENTS

Reactive intermediates such as cations are frequently used to estimate the energies of transition states. Molecular modeling can be used to compare different intermediates; the results are only meaningful when the electronic differences between the two intermediates are small so that the primary differences are due to sterics. In the case of the Friedel-Crafts alkylation, the cations resulting from attack at either the *ortho* or *para* position should have approximately the same amount of electronic stabilization (this can be confirmed by writing out the sequence of resonance structures associated with the cations). Therefore, any energy differences between the two cations should be due to steric interactions.

When an sp^2 carbon adjacent to a bulky group such as the *t*-butyl group is attacked by an electrophile, there is considerable relief of strain as the carbon being attacked goes from sp^2 to sp^3 hybridization. However, as the reaction proceeds to the product, the strain is reintroduced as the carbon goes back to sp^2 hybridization, forcing the methyl and the *t*-butyl to occupy the same plane again. In the case of toluene, there is much less strain involved, as the adjacent groups are now two methyls rather than one methyl and one *t*-butyl.

To calculate the ratio of products in a given reaction, it is important to know if the reaction is under kinetic control or under thermodynamic control. Kinetic control means that the products that are formed fastest will be present in the greatest amount. Product stability (a thermodynamic property) will have no direct bearing on the product distribution. Instead, the differences in activation energies for the various reaction pathways would allow us to predict the product distribution. Basic molecular-modeling techniques do not allow us to estimate the energy of the transition states; however, we can estimate the relative amounts of the intermediate. If there is little difference in activation energy of the various pathways, then the relative amounts of the various carbocations will give us a good approximation of the product distribution. If, on the other hand, there is a large difference in activation energy of the various pathways, then the energy differences of the carbocations may not accurately reflect the ratio of products.

In the Friedel-Crafts reactions considered in this problem, it should be noted that statistical considerations alone (in the absence of either electronic or steric factors) would predict twice as much *ortho* product as *para* product, since there are two *ortho* positions that could be attacked.

INSTRUCTIONS

Consider reactions **a** and **b**, shown below.

1. Draw and minimize the four products. Record the minimized energies.

2. Draw and minimize the four carbocation intermediates. (Note that it is possible to explicitly indicate a cation in ChemDraw; be sure to do this before copying the structure over to Chem3D. You will know that you have done this correctly if, in Chem3D, when placing the cursor over the cation carbon, the pop-up box indicates C-Cation.) Record the minimized energies.

How to Draw in the ChemDraw Window

Click on the Benzene Ring tool in the toolbar (it looks like a benzene ring, big surprise). Then click in the drawing space. You will get a benzene ring. To add an alkyl substituent, click on the Solid Bond tool (it looks like a straight line on the toolbar), and then put the cursor over one of the carbons on the benzene ring; you should see a "hit box" (on my computer it shows up as green). With the hit box active, click. You should sprout a bond off of the benzene ring, giving you toluene.

To draw the *ortho* cation intermediate, first use the Eraser tool to eliminate the double bond between the *ortho* and *meta* carbons. Add a methyl group to the *ortho* carbon. Then assign a + charge to the *meta* carbon by choosing the + charge from the Chemical Symbols drop-down list on the toolbar, placing the cursor over the *meta* carbon (activating the hit box), and clicking. You should see a + (inside a circle) appear next to that carbon.

How to Minimize Your Drawing and Get Its Energy

Click on Calculations, MM2, Minimize Energy, Run. Your final energy will appear at the bottom of the screen.

QUESTIONS

1. For reaction **a**, what is the energy difference between the two cations?

2. For reaction **a**, what is the energy difference between the two products?

3. For reaction **b**, what is the energy difference between the two cations?

4. For reaction **b**, what is the energy difference between the two products?

5. Estimate the relative amounts of the products that are formed in reactions **a** and **b**, using the energy differences of both the cations and the products as predictors. Which is a better estimate? Why?

REPORT

Your report should include the following:

1. Title page

2. Abstract

3. Data (structures and corresponding energies)

4. Answers to problems, including calculations

5. Conclusions

An Aerobic Alcohol Oxidation Using a Copper(I)/TEMPO Catalyst—Prelab

Prelab Report: Due in the beginning of the lab period.

Name (print): ____

Lab Section (Circle One): Mon   Tues   Wed   Thur   Fri   AM/PM

1. Define a chemical catalyst.

2. What two main points make this a "green" reaction?

3. What is the first substance to get oxidized in the reaction?

4. How is color used as an indicator in this reaction?

5. After you transfer your reaction mixture to a separatory funnel, add water and pentane and shake, you will see two layers in your funnel. Which layer should be the top and which should be the bottom layers?

6. Please fill this table completely:

Substance	M. Wt. (g/mol)	d (g/mL)	Molar equiv.	Amount
Water				
Pentane				
Acetone				15 mL
TEMPO				40 mg
Bipyridine				
CuBr				
NMI			approx	4 drops
Starting alcohol: (4-bromophenyl)methanol				
Product aldehyde: 4-bromobenzaldehyde				

An Aerobic Alcohol Oxidation Using a Copper(I)/TEMPO Catalyst

BACKGROUND

Comparisons to other oxidants and stoichiometry

No doubt that you have heard of many metal-based oxidizing agents, such as pyridinium chlorochromate (PCC), CrO_3 in conc. H_2SO_4 (Jones reagent), $KMnO_4$ and OsO_4 in our Organic Chemistry lecture. Such reagents are traditionally used in organic synthesis research labs to accomplish a range of oxidations; however, despite their utility, they feature a number of disadvantages:

a.   The reagents are often highly toxic, environmentally harmful and difficult to dispose;

b.   They are typically used in stoichiometric amounts (i.e., the same molar amount as the substrate that is being oxidized);

c.   They yield by-products (such as inorganic salts) that have to be separated from the desired product;

d.   They often promote overoxidation, in which the desired product undergoes further oxidation before it can be isolated. For example, aldehydes are commonly overoxidized to carboxylic acids.

These and other factors often make it difficult to scale-up stoichiometric metal-based oxidation reactions for use in the synthesis of fine-chemicals and pharmaceuticals.

Instead of using **stoichiometric** oxidants alone, many researchers focus on the development of **catalytic** systems that enable the use of atmospheric O_2 as the stoichiometric oxygen source. (Recall that a catalyst is an additive used in a substoichiometric amount that increases the rate of a chemical reaction without itself being consumed or undergoing a net chemical change.) In other words, a catalyst lowers the activation energy of a chemical process by lowering the energy of the transition state (ΔG) for conversion of reactants to products.

So, why is this reaction a "green" reaction? We will be using substoichiometric amounts of catalyst reagent, and air as the "oxidant," so we will be leaving this reaction open to the atmosphere. This reaction is relatively safe to be run on the laboratory bench-top, if need, though the preferred choice is to run this in a negative-atmosphere environment, such as a fume hood.

Mechanistic discussion

In this experiment you will use a catalytic system consisting of copper(I) bromide (CuBr), 2,2'-bipyridine (bpy), N-methyl imidazole (NMI), and a stable organic radical (TEMPO, 2,2,6,6-**te**tramethyl-1-**p**iperidinyloxyl) in combination with atmospheric oxygen as the stoichiometric oxidant to convert a benzyl alcohol in this case (4-bromophenyl)methanol, to the corresponding aldehyde. You will determine if the reaction was a success by comparing the starting material IR to the final product IR.

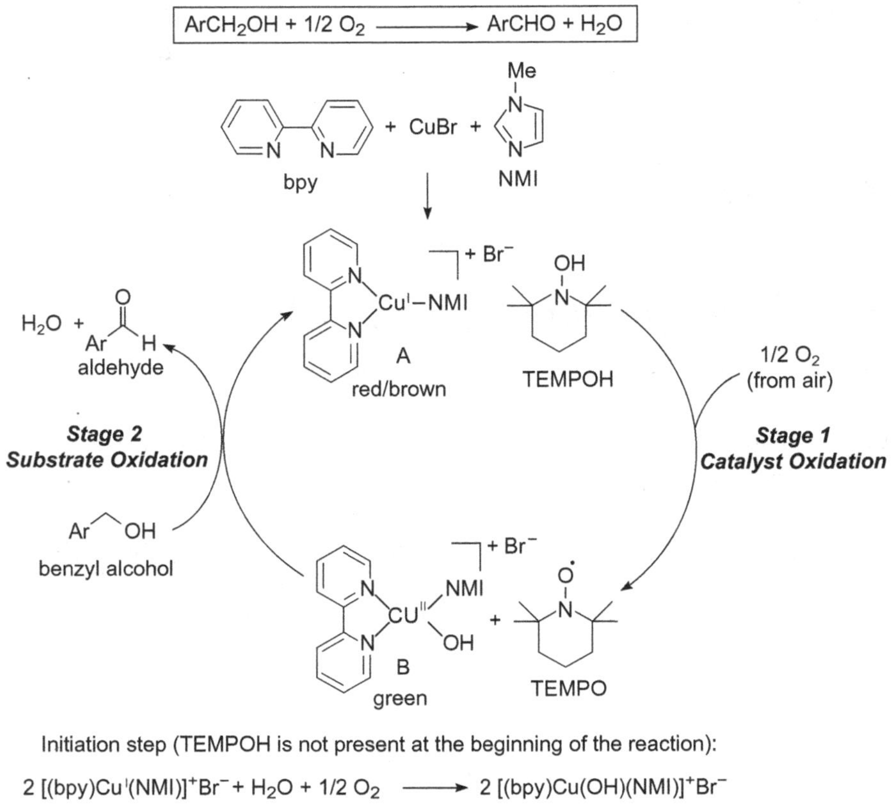

Scheme 1—*Simplified version of the catalytic cycle of the Cu(I)/TEMPO oxidation of benzyl alcohol.*

From *Journal Of Chemical Education* by Nicholas J. Hill, Jessica M. Hoover and Shannon S. Stahl. Copyright © 2012 by The American Chemical Society and Division of Chemical Education, Inc. Reprinted by permission. http://pubs.acs.org/doi/full/10.1021/ed300368q

Mechanistic details of this oxidation are beyond the scope of an introductory course, but suffice it to say that Hoover, Ryland, and Stahl present compelling evidence that support a two-stage catalytic mechanism consisting of (Stage 1, above) "catalyst oxidation" in which Cu(I) and TEMPOH are oxidized by O$_2$ via a binuclear Cu$_2$O$_2$ intermediate and (Stage 2, above) "substrate oxidation" mediated by Cu(II) and the nitroxyl radical of TEMPO *via* a Cu(II)-alkoxide intermediate. The substrate oxidation sequence can be presented as a pair of simplified electron-pushing mechanisms (Schemes 2 and 3). The alcohol substrate can react with the Cu(II)–hydroxide complex **B** to form the Cu(II)–alkoxide intermediate **C** with the subsequent release of water (Scheme 2). This reaction corresponds to a Lewis acid–base-like mediated reaction. The Cu(II) bound alkoxide of complex **C** can then react with the TEMPO radical in a radical-like mechanism to afford the aldehyde product (Scheme 3). The Cu(II) center is reduced to Cu(I) in this step, resulting in the formation of complex **A** together with TEMPOH. Overall, the conversion of an alcohol to an aldehyde is a net two-electron oxidation reaction.

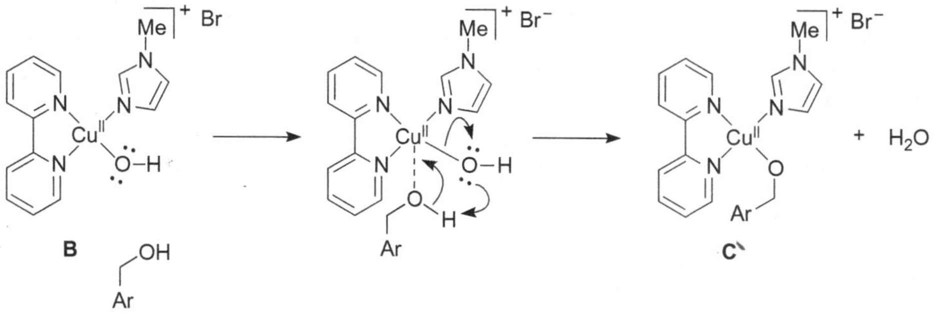

Scheme 2—*Electron Pushing Mechanism for Formation of the Cu^{II}—Alkoxide Intermediate C.*

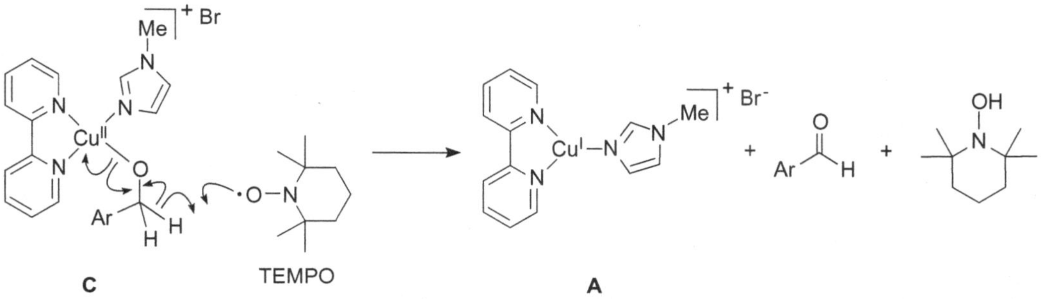

Scheme 3—*Electron Pushing Mechanism for Formation of the Aldehyde Product from the Reaction of Cu^{II}—Alkoxide Complex C and TEMPO.*

PROCEDURE

Instructor and Lab assistants: Set out a warming plate in a fume hood and set it on low to aid in student solvent evaporation. Please set up melting point apparati, IR Spectrometers, and TLC chambers (1:4 EtOAc/ Hexane) and lamps.

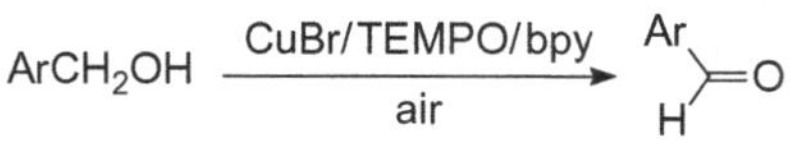

$$ArCH_2OH \xrightarrow[\text{air}]{\text{CuBr/TEMPO/bpy}} Ar\text{-CHO}$$

Scheme 4—*General reaction equation of the Cu(I)/TEMPO oxidation of aryl alcohols.*

Work in pairs, and in a fume hood, if available. To a 125 mL Erlenmeyer flask containing a stir bar add the (4-bromophenyl) methanol (2.5 mmol) and acetone (15 mL). Weigh CuBr (35 mg), bpy (40 mg), and TEMPO (40 mg) into *separate* small glass vials.

To the rapidly stirred solution of the unknown alcohol add sequentially CuBr, bpy, and TEMPO as solids, following each addition with ~3 mL rinse of solvent (for total of ~25 mL solvent volume). Add NMI (4 drops, the color of the reaction solution will fade) and stir the mixture rapidly until you observe a significant color change, then stir for a further 5 min. What color does it change from/ to?

Transfer the reaction mixture to a 250 mL separatory funnel. Add water (25 mL) and pentane (25 mL) and extract the organic product into the organic solvent with vigorous shaking. Remove the aqueous layer (think about which layer is which!) and extract it with pentane (10 mL). Combine BOTH the organic layers and stir over anhydrous $MgSO_4$ for ~10 min.

Remove the $MgSO_4$ by filtration (fluted paper funnel) and transfer the filtrate into a tared beaker or flask. Evaporate the solvent in a fume hood using warming plate or under a stream of dry nitrogen. After the solvent is removed, note the appearance of the product.

Determine the % yield first by weighing the tared flask, THEN transfer the solid to a clean, dry screw-top vial, and completely label it. Take a sample and analyze it by **melting point** and **IR** and **TLC** (1:4 EtOAc/Hexane). Compare the R_f of the starting material and final product R_f. Report the TLC data to two decimal places and sketch out your TLC in your lab report. Compare your product IR to the attached starting material IR and assign all important peaks to BOTH the starting material AND your product IRs. Include this with your lab report. Hand in the product along with the other products when those are due.

QUESTIONS

In addition to your normal lab report write-up, experimental data, and the above IR assignments, please include the answers to these questions and requests:

1. Please sketch out the theoretical ^{1}H-NMR spectrum for your product. Label the scale, assign shifts, multiplicity, and integration.

2. What evidence would appear in the product IR spectra indicating over-oxidization of the aldehyde?

LAB REPORT

Lab report incorporates the full procedure and data, and synthetic scheme that includes the exact benzyl alcohol used. Do not forget percent yield calculations, melting point, TLC data as described above, IR of product and starting material (provided below) with peaks label, and the answers to the questions above. The report should be scanned and saved as a single document in PDF format. You may submit it via e-mail or print it out depending on what your instructor wants.

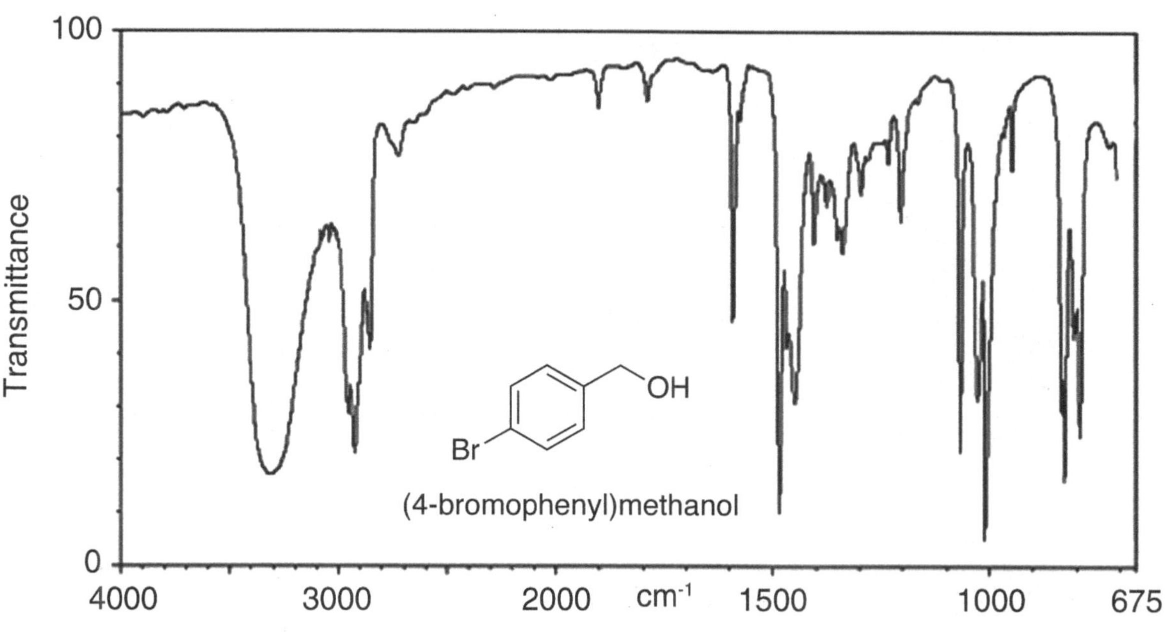

Figure 1—*(4-Bromophenyl) methanol starting material IR.*

Adapted from Chemical Book, 2016

REFERENCES

J. Am. Chem. Soc. 2011, 133, 16901–16910

J. Chem. Educ. 2013, 90, 102–105

Preparation of Benzil—Prelab

Prelab Report: Due at the Beginning of the Lab Period

Name ____

Lab Section (Circle One): Mon   Tues   Wed   Thur   Fri   AM/PM

1. What oxidizing agent is used in this reaction?

2. What color is the reaction originally? What color does it turn?

3. What solvent will you use to recrystallize the crude benzil?

4. Provide the requested physical properties:

   Melting point, benzoin:

   Molecular weight, benzoin:

Melting point, benzil:

Molecular weight, benzil:

Bolling point, benzoin:

Density, ethanol:

PREPARATION OF BENZIL

BACKGROUND

Traditionally, the oxidation of benzoin to benzil has been performed using nitric acid as the oxidizing agent. This method usually yielded product still contaminated with unreacted benzoin, and the use of concentrated nitric acid certainly was not "green." In this experiment, we will use the mild oxidizing agent copper(II) acetate in an aqueous ammonium nitrate solution.

The Cu(II) ion oxidizes the benzoin while it is reduced to the Cu(I) ion [eqn 1]. Then the ammonium nitrate oxidizes the newly made Cu(I) back to Cu(II), while the nitrate ion is reduced to the nitrite ion [eqn 2]. So in this system, the Cu(II) ion is a catalyst and is continually remade as the benzoin to benzil oxidation progresses.

Equation 1

$$2 \, H^+ + 2 \, Cu^+ + NH_4NO_3 \longrightarrow 2 \, Cu^{+2} + NH_4NO_2 + H_2O$$

Equation 2

Ammonium nitrite, in the presence of acid, decomposes to give nitrogen gas and water [eqn 3].

$$NH_4NO_2 \, (\text{in acid}) \longrightarrow N_2 + 2 \, H_2O$$

Equation 3

PROCEDURE

1. Place 0.30 g of benzoin in a 5-mL conical reaction vial with a magnetic spin vane. Add 2 mL of the $Cu(OAc)_2$ solution—which should be a bluish color— and attach an air condenser. Heat with stirring on a hot plate, using the aluminum heating block until the solids dissolve completely and the mixture turns green.[1] This should take two to three minutes. Continue to heat the solution at a gentle reflux for sixty minutes.

2. When the reflux period is ended, allow the reaction to cool for a few minutes; then remove the air condenser. Pour the contents of the reaction vial into a small beaker containing 4 mL of cold water. Rinse the vial and spin vane with an additional 1 mL of cold water, and add this to the beaker as well. Stir the mixture well. At this point, you should have a yellowish solid. Collect the solid by vacuum filtration, and wash it well with a further 5 mL of cold water. Continue to pull air through the crude solid for about five minutes.

3. Weigh the crude benzil and place it in a 10-mL Erlenmeyer flask. Recrystallize from 95% ethanol (the solubility of benzil in ethanol is approximately 1 g/7 mL). Collect the yellow needles by vacuum filtration, and allow to dry in your laboratory drawer at least overnight.

4. Obtain the weight, melting point, and IR spectrum of your recrystallized benzil. Calculate the percent yield. (The melting point of pure benzil is 95°C; your product will likely melt about 92 to 94°C.) Save the product for later conversion to tetraphenylcyclopentadienone.

---

[1] If your starting benzoin is not terribly pure, the reaction mixture may turn more of a light tan or brownish color. If the color does not turn to green within thirty minutes, add an additional 1 mL of the $Cu(OAc)_2$ solution, and continue heating for thirty minutes more.

PREPARATION OF TRIPHENYLMETHANOL—PRELAB

Prelab Report: Due at the Beginning of the Lab Period

Name _______________________________________

Lab Section (Circle One): Mon   Tues   Wed   Thur   Fri   AM/PM

1. How does the alkyl portion of a Grignard reagent behave?

2. Draw the specific Grignard reagent you will prepare (bond-line notation, please!).

3. Show the mechanism for reaction of your Grignard reagent with benzophenone.

4. What is the principal impurity formed in this reaction, and how will you separate it from the desired product?

5. Why must the glassware and ether be completely anhydrous? (HINT: What happens to a Grignard reagent in the presence of water?)

6. Provide the requested physical properties:

Density, diethyl ether: Boiling point, diethyl ether:

Molecular weight, triphenylmethanol: Melting point, triphenylmethanol:

Density, bromobenzene: Boiling point, bromobenzene:

PREPARATION OF TRIPHENYLMETHANOL

INTRODUCTION

Organometallic compounds contain a carbon-metal bond. One example is organomagnesium compounds (typically called Grignard reagents, after Victor Grignard[1] who discovered them in 1901), which are prepared by reacting an organic halide with magnesium metal in an anhydrous ethereal solvent. The ether solvent plays a crucial role in the reaction by forming a soluble complex with the Grignard reagent. Grignard reagents can be formed with primary, secondary, and tertiary alkyl halides, as well as with aryl and vinyl halides, and they can then be subsequently reacted with a variety of classes of organic compounds to form a new carbon-carbon bond. The carbon-magnesium covalent bond is polar, with a partial negative charge on the carbon, because the metal is less electronegative than carbon. The carbon is nucleophilic and reacts in further reactions as if it was a carbanion. Since the Grignard reagent is also a strong base, it will react with any compound having an acidic proton; water, alcohols, terminal alkynes, phenols, and carboxylic acids will destroy the reagent. However, Grignard reagents will react with polar π-bonded functional groups (like the carbonyl group of aldehydes, ketones, and esters) to add the partially negative carbon of the carbon-magnesium bond to the partially positive carbon of the carbonyl. The initial product of this type of reaction would be an alkoxide, which is then hydrolyzed to an alcohol.

In this experiment, the Grignard reagent (phenylmagnesium bromide) is reacted with a ketone (benzophenone) to initially give the Grignard adduct, which is hydrolyzed to the tertiary alcohol, triphenylmethanol.

Some coupling between the Grignard reagent and unreacted aryl halide leads to biphenyl, the major side product (which must be removed in the final purification of the alcohol).

PROCEDURE[2]

It is essential that all glassware used be clean and scrupulously dried. Clean the following items, place them into a labeled beaker, and give them to your TA <u>during the laboratory period before this experiment:</u> two 5-mL reaction vials, 50-mL Erlenmeyer flask, Claisen head, drying tube, and a glass rod (REMOVE all plastic caps or rubber O-rings!). Our laboratory staff will be responsible for drying your glassware in the 120°C drying oven overnight; previously dried 25-mL round-bottom flasks, spin bars, and Pasteur pipets will be available when you do the experiment.

Go to the drying oven, and obtain all the necessary materials (CAREFUL: They will be at 120°C, which is HOT). As soon as the glassware is cool enough to handle, first charge the drying tube with anhydrous calcium chloride pellets (held in place by cotton plugs), and then assemble the remainder of the apparatus as pictured, except for the rubber septum (see Figure 1). Obtain 0.15 g of magnesium turnings,[3] and place them (along with the stir bar) into the round-bottom flask. Place the rubber septum over the open port of the Claisen head.

Courtesy of Kenneth F. Cerny

Figure 1 Grignard setup

Take your 50-mL Erlenmeyer flask to the hood, and obtain about 20 mL of anhydrous ethyl ether. Stopper the flask, and use this as your source of dry ether throughout the experiment, transferring the ether with a dried Pasteur pipet (keep the flask stoppered when not actually transferring ether). Place about 0.70 mL of bromobenzene into one of the oven-dried, preweighed 5-mL reaction vials, and then reweigh it to determine the weight of bromobenzene actually taken. Add 4 mL of dry ether to the vial, and swirl to dissolve. After the bromobenzene dissolves, withdraw about 1 mL of this solution into the syringe, cap the vial, and insert the syringe needle through the rubber septum. Add the bromobenzene solution (that is in the syringe!) to the magnesium in the round-bottom flask. Be sure that the round-bottom flask is above the center of your stir-hot plate (about 60°C), and begin to stir the mixture very gently to avoid moving the magnesium onto the sides of the flask (try to get the stir bar to "wiggle" instead of spin). Observe the reaction mixture closely; the evolution of bubbles from the magnesium surface and/or cloudiness indicates that the reaction has started. It may be necessary to heat the mixture to start the reaction (ether has a low boiling point, so just lowering the apparatus closer to the hot plate should begin the reflux of the ether). Once the mixture starts to boil, however, remove it from the heat by raising the apparatus away from the hot plate. If the reaction is indeed occurring, the mixture will remain boiling on its own without the added heat from the hot plate.

If the reaction does not start within five minutes, detach the flask momentarily, hold it in the palm of your hand, and <u>carefully</u> (do not punch a hole in the bottom of the flask!) crush the magnesium pieces with the end of the dry glass stirring rod for at least thirty seconds. Reattach the flask to the reaction apparatus, and heat the mixture gently to start the reaction. You may need to repeat this crushing and reheating procedure in order to start the reaction. If the reaction does not start within about ten minutes, consult your instructor.[4]

As the reaction continues, you should observe the solution becoming cloudy (there may also be a color change to a muddy brownish-yellow). Continue the addition of your bromobenzene-ether solution by refilling the syringe and adding it to the flask over a period of about twelve minutes. Ideally the reaction will continue to boil on its own as you slowly add the remainder of this solution. But you may need to heat the mixture to boiling if the reaction subsides. The magnesium metal will gradually disintegrate (although it should not completely disappear). After the addition of the bromobenzene-ether solution is complete, the reaction mixture should be heated to reflux for about fifteen minutes. (Watch the level of liquid in the round-bottom flask closely; do not allow all the ether to evaporate!) During this reflux time, rinse the original bromobenzene-ether reaction vial with about 2 mL of anhydrous ether. Draw this solution into the syringe, and add it to the reaction flask as well. After the fifteen-minute reflux, allow the reaction mixture to cool, but continue stirring.

During the above fifteen-minute reflux period, dissolve 1.09 g of benzophenone in 2 mL of anhydrous ether in the other oven-dried 5-mL reaction vial. Use a cap to close this vial until you are ready for the benzophenone solution. Once your reaction mixture (the Grignard reagent) is cooled appreciably below boiling, add the benzophenone-ether solution (again using the syringe). The rate of this addition should be such that the main reaction mixture does not boil. Rinse the benzophenone-ether vial with 1 mL of anhydrous ether, and again use the syringe to add this solution to the reaction flask. Continue stirring as long as possible (as the mixture cools, it will turn a bright pinkish-red and solidify; it will then turn white and the magnet stirrer will be stuck in the solid mass). You may now remove the reaction flask from the apparatus and occasionally stir the contents with a glass stirring rod. Keep the flask capped when you are not stirring (the formation of the Grignard adduct will be complete after about fifteen minutes). **<u>You MUST get to this point in your first laboratory period!</u>** Disassemble the reaction apparatus, cap your reaction flask, and seal it with parafilm until the next laboratory period.

Obtain about 6 mL of 6 M hydrochloric acid, and begin to add this solution dropwise to the reaction mixture contained in the round-bottom flask. Again stir the mixture with a stirring rod. The hydrolysis of the Grignard adduct is exothermic, so some ether will be lost by evaporation. As you continue to add the dilute hydrochloric acid, monitor the volume of the top ether layer. (Try to maintain about 10 mL of ether. You will need to add some ether periodically. Now the ether does not need to be anhydrous.) Continue to add the acid and ether until two distinct layers are evident. No solid should be present (you may need to break up solid chunks with a stirring rod; occasionally some magnesium remains).

Remove the stir bar, and transfer both layers into a small separatory funnel (leaving any magnesium behind). Rinse the flask with 1 mL of ether (not anhydrous) and transfer the ether solution to the separatory funnel. If some white solid forms, add more ether to dissolve it. The upper ether layer contains the triphenylmethanol, while the lower aqueous layer contains the excess hydrochloric acid and inorganic salts. Separate and save the layers. Extract the water layer with 5 mL of ether. Separate and save the layers. Combine the ether layers into a small Erlenmeyer flask, and dry the ethereal solution over granular anhydrous sodium sulfate by allowing the flask to sit for at least fifteen minutes. Separate the dry ether solution from the solid drying agent by decanting it into a small beaker. Evaporate the ether in the hood, leaving an oily crude solid mixture (triphenylmethanol and biphenyl). Add 3 mL of petroleum ether (30–60°C)[5] and heat slightly with stirring to dissolve the biphenyl. Cool the mixture to room temperature, and collect the solid triphenylmethanol by vacuum filtration using a Hirsch funnel. Wash the collected solid with 0.5 mL of petroleum ether, and then draw air through the funnel for ten minutes. Weigh your crude product.

Recrystallize your product from hot isopropanol. After cooling in an ice bath, collect the purified solid product by vacuum filtration using a Hirsch funnel. Wash the solid with ice-cold isopropanol, draw air through the funnel for ten minutes, and then allow the product to air-dry until the next laboratory period.

Weigh your purified product and calculate the percent yield. Take the melting point and an infrared spectrum of the purified triphenylmethanol. Submit your product in a properly labeled sample bottle to your instructor.

[1] He received the Nobel Prize in 1912 for his work in organic synthesis with organomagnesium compounds.

[2] Since the Grignard reagent is so reactive and cannot be stored, this experiment must be completed at least to the addition of the benzophenone solution. ·

[3] The magnesium surface must be shiny! You may need to scrape the surface to remove the oxide coating. Avoid touching the magnesium with your fingers.

[4] There are other possible ways to start the reaction: adding an iodine crystal or preparing a small amount of the Grignard reagent in a test tube and adding this to the reaction flask.

[5] Petroleum ether is a hydrocarbon mixture that will dissolve biphenyl (also a hydrocarbon) and leave behind the alcohol, triphenylmethanol. The fact that a hydrocarbon mixture is called an "ether" is a misnomer.

SEPARATION OF A MIXTURE— PRELAB

Prelab Report: Due at the Beginning of the Lab Period

Name ____

Lab Section (Circle One): Mon   Tues   Wed   Thur   Fri   AM/PM

1. Briefly explain how to make a saturated solution of sodium bicarbonate.

2. When the sodium bicarbonate solution is added to the salicylic acid/naphthalene mixture dissolved in ether, a gas is evolved. What gas is this?

3. How do you tell when a solution is acidic?

4. What is the purpose of the sublimation and recrystallization steps?

5. What solvent is used to recrystallize the salicylic acid?

6. Provide the literature values for the requested physical properties.

Melting point of salicylic acid:

Density of diethyl ether:

Molecular weight of naphthalene:

Melting point of naphthalene:

Separation of a Mixture

PROCEDURE

Extraction

1. A 0.70 g sample of the salicylic acid–naphthalene mixture is dissolved in 10 mL of diethyl ether. This solution is placed in a small-to mid-volume separatory funnel. Cautiously add 10 mL of a saturated aqueous sodium bicarbonate solution to the separatory funnel (gas evolution is observed—what gas is being formed?). After the initial gas evolution has subsided, the separatory funnel is stoppered and shaken, venting frequently. (Your instructor will demonstrate the proper way to use a separatory funnel.) Allow the two layers to separate. Remove and save the aqueous layer (which layer is that— top or bottom?) using a Pasteur pipet. Add a second 10-mL portion of saturated aqueous sodium bicarbonate to the ether layer in the test tube, and repeat the extraction.

2. Combine the two aqueous extractions in a beaker, and cautiously acidify with concentrated HCl to precipitate the salicylic acid. Collect the crude salicylic acid by vacuum filtration (Hirsch funnel) and allow to dry in an open container until the next laboratory period.

3. Add 4 mL of water to the ether layer in the separatory funnel, and shake well to remove any leftover sodium bicarbonate. Separate the layers, and place the ethereal layer in a labeled erlenmeyer flask. Add a small amount of drying agent to that erlenmeyer to remove the last traces of water. Allow the ether to dry for at least 10 min. Then transfer the ether layer to a clean, dry beaker (be sure to leave the drying agent behind in the erlenmeyer) using a Pasteur pipet. Place the beaker in the hood, and allow the solvent to evaporate; then store the crude naphthalene in a covered container (place a watch glass or some Parafilm over the beaker) until the next laboratory period. Make sure everything is labeled properly!

Sublimation

1. Determine the weight and melting point of the crude naphthalene, and then place it in a 25-mL filter flask. Use the neoprene adapter to suspend a centrifuge tube filled with ice over the solid in the flask (this acts as a "cold finger"). Make sure that the bottom of the centrifuge tube is no more than a half inch from the bottom of the filter flask. It is a good idea to wrap a Kimwipe or two around the top of the neoprene adapter to catch any condensation that takes place on the outside of the centrifuge tube.

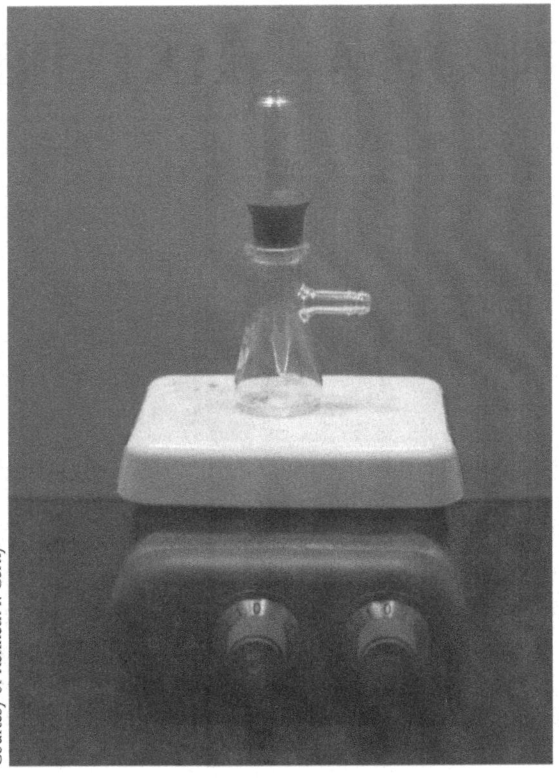

2. Gently warm the naphthalene (no vacuum!) using a low setting on the hot plate. The naphthalene will sublime and be trapped on the outside of the cold finger (and possibly on the walls of the filter flask as well; this is okay). If it starts "snowing" inside the flask, turn off the heat and allow the apparatus to cool. Then carefully remove the cold finger, and scrape the purified naphthalene onto a piece of weighing paper. Be sure to invert a beaker or something over the naphthalene, as a stray gust of air can easily blow it away! Replace the cold finger (add more ice if necessary), and continue the sublimation. When all of the crude naphthalene has sublimed (there may be discolored material left in the bottom of the filter flask), determine the weight and melting point of the purified naphthalene, and place it in a labeled vial to be turned in later.

Recrystallization

1. Determine the weight and melting point of the crude salicylic acid. Recrystallize the salicylic acid. The important fact to keep in mind whenever doing a recrystallization is that you must use a <u>minimum amount of hot, boiling solvent</u> to dissolve the solid being recrystallized. In this case, you will use approximately 0.5 mL of water. Once the solid is dissolved, allow it to cool first on the bench top and then in an ice bath. Crystals should form. Collect the crystals by vacuum filtration.

2. Allow the recrystallized salicylic acid to dry in an open container until the next laboratory period. Then determine its weight and melting point, and place it in a labeled vial to be turned in later.

DATA TO BE DETERMINED

You should record five melting points:

1. The melting point of the naphthalene/salicylic acid mixture

2. . The melting point of the crude naphthalene

3. The melting point of the crude salicylic acid

4. The melting point of the pure naphthalene

5. The melting point of the pure salicylic acid

You should also calculate a percent recovery for both the salicylic acid and the naphthalene. To calculate percent recovery, assume that the salicylic acid–naphthalene mixture was a 1:1 mixture by weight. Therefore, a 2 g sample would consist of 1 g naphthalene and 1 g salicylic acid. The percent recovery is the amount you actually recovered, divided by the amount you should have recovered, times 100%.

For example, assume you recovered 0.63 g of naphthalene from a 2.2 g sample of the mixture:

$$\textbf{Percent Recovery (naphthalene)} = \frac{\textbf{0.63 g recovered}}{\textbf{1.1 g present}} \times 100\% = 57\% \text{ recovery}$$

STEREOSELECTIVE REDUCTION OF CAMPHOR—PRELAB

Prelab Report: Due at the Beginning of the Lab Period

Name ____

Lab Section (Circle One): Mon   Tues   Wed   Thur   Fri   AM/PM

1. How can you determine if your reduction was successful?

2. What important side reactions can occur in this experiment?

3. The NMR spectrum of isoborneol shows three separate methyl peaks, but in the NMR spectrum of borneol the three different methyl groups are not resolved (that is, they appear at the same chemical shift). Explain.

4. Provide the requested physical properties:

    Melting point, camphor:

    Molecular weight, camphor:

    Boiling point, methanol:

Stereoselective Reduction of Camphor

Camphor → (NaBH₄ / CH₃OH) → Isoborneol + Borneol

INTRODUCTION

The nucleophilic addition of a hydride ion [H⁻] to the carbonyl group (followed by hydrolysis) results in the reduction of aldehydes to primary alcohols and ketones to secondary alcohols. Common reducing agents used in the organic lab are sodium borohydride [$NaBH_4$] and lithium aluminum hydride [$LiAlH_4$]. The hydride ion is a powerful base as well as a nucleophile. In this experiment, we will use sodium borohydride, since it is easier to handle. (The reduced reactivity of sodium borohydride allows it to be used in alcohol solvents, whereas lithium aluminum hydride reacts violently with hydroxylic solvents to produce hydrogen gas.)

The reasons for stereochemical control in reduction reactions are still not totally understood. There are at least three possible influences: (1) steric approach of the nucleophile, (2) product stability (thermodynamics), and (3) electronic influence of substituents. For camphor and other bicyclic systems an analysis of the steric approach of the hydride ion explains the isomeric product percentages. If the camphor carbonyl is regarded as flat, the approaching HB ion can add to the top or bottom face of the carbonyl—however, not with equal ease! If one examines the camphor molecule, the approach of the hydride ion from the bottom side (endo-approach) is easier than the approach from the top (exo-approach), since a large steric interaction is apparent with one of the two geminal methyl groups. The relative amounts of the two products in the reaction mixture can be determined by spectroscopic techniques—we will use ¹H NMR.

PROCEDURE

Synthesis

1. In a 10-mL (or 25-mL) Erlenmeyer flask, dissolve 0.10 g of camphor in 0.5 mL of methanol.

    Warning! Take extreme care when handling sodium borohydride; it will react violently if it comes in contact with water—flammable hazard!

2. In portions, cautiously and slowly, add 0.060 g of sodium borohydride to the methanol solution. If the additions appear violent and the flask temperature elevates above room temperature, place the flask in an ice-water bath (carefully—be sure no water gets into the flask!). After all the borohydride has been added, heat the contents to a <u>gentle</u> boil on a hot plate for about three minutes. (NOTE: Take care that the contents do not evaporate! If you observe the volume getting smaller, just add 0.5 mL of methanol.)

3. After heating, allow the reaction mixture to cool to room temperature. Then carefully add 4 mL of ice water. Collect the white solid by vacuum filtration using a Hirsch funnel. Allow the solid to dry several minutes with the suction applied. Transfer the solid product to a 25-mL Erlenmeyer flask. Add 5 mL of diethyl ether to dissolve the product, and then add anhydrous sodium sulfate to dry the ethereal solution.

4. Remove the ether solution using a Pasteur pipet, and place it into a <u>dry</u>, <u>preweighed</u>, clean, small beaker. To recover more product, add 1 mL of diethyl ether to the flask containing the drying agent and swirl the flask. Again remove the ether solution, and combine the two ether solutions in the preweighed beaker. Place the beaker in the hood until the ether has evaporated, yielding an off-white solid product mixture. Reweigh your beaker (and solid), and calculate both your product weight and percent yield. Determine the melting point of your product mixture (racemic isoborneol melts at 212°C; racemic borneol melts at 206–207°C). Obtain the infrared spectrum of the product mixture, and determine if you were successful in reducing camphor.

^{1}H NMR Analysis of the Product Mixture

The hydrogen on the carbon bearing the hydroxyl group appears at approximately 4 ppm for borneol and 3.6 ppm for isoborneol. You can obtain the product ratio by integrating these peaks in the ^{1}H NMR spectrum of the product mixture after reduction. A sample copy of this spectrum is provided below (with an expansion of the 3.2–4.2 ppm range). Determine the percentages of borneol and isoborneol from the integration curves as such:

$$\frac{\text{integration (isoborneol)}}{\text{integration (total of both peaks)}} \times 100\% = \% \text{ isoborneol}$$

QUESTIONS

1. What is the mechanism for the formation of these alcohols starting from camphor?

2. A student product from this experiment was analyzed by IR and showed a band at 1755 cm^{-1}. Was this an expected band (peak) in the product? Why is the peak in the IR spectrum?

3. How do you rationalize the percentages of the isomeric alcohols found in your reaction product mixture? Use theory to explain your results.

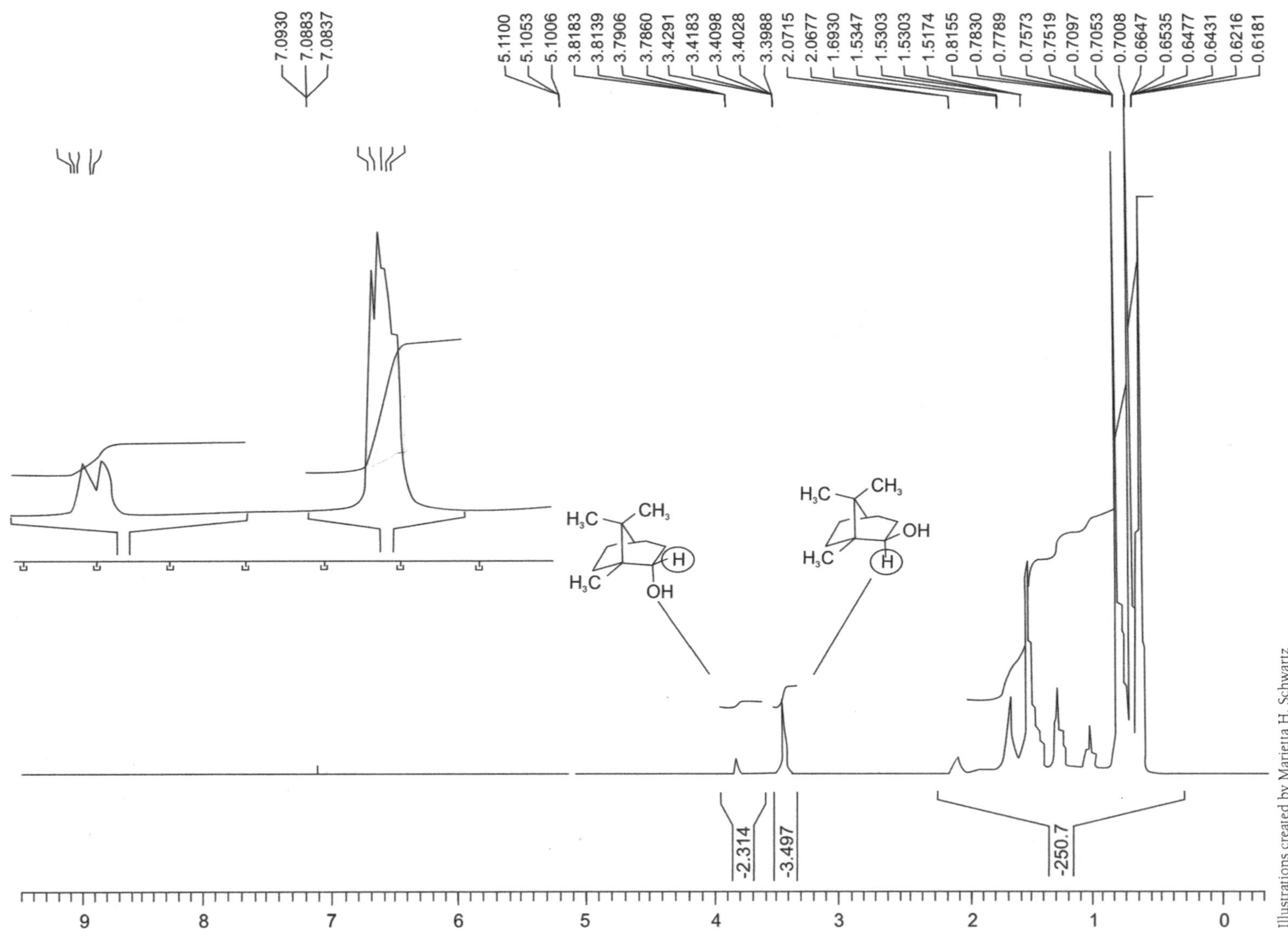

Illustrations created by Marietta H. Schwartz

Tetraphenylcyclopentadienone—Prelab

Prelab Report: Due at the Beginning of the Lab Period

Name ____

Lab Section (Circle One): Mon   Tues   Wed   Thur   Fri   AM/PM

1. What color should your product be?

2. What is the name reaction utilized in this experiment?

3. What is the recrystallization solvent that can be used to purify the product?

4. What functional groups are present in your product?

5. What technique will you use to collect your product?

6. If you started with 0.274 g of benzil and 0.198 g of dibenzyl ketone:

    a.   What is the limiting reagent?

    b.   What is the theoretical yield of tetraphenylcyclopentadienone?

7. Provide the requested physical properties:

Melting point, dibenzyl ketone:

Molecular weight, dibenzyl ketone:

Melting point, tetraphenylcyclopentadienone:

Molecular weight, tetraphenylcyclopentadienone:

TETRAPHENYLCYCLOPENTADIENONE

This is the third step in the synthetic reaction sequence outlined earlier in this laboratory manual. Be sure to save your product for use in the final step of the synthetic sequence!

REACTION

PROCEDURE

First Laboratory Period

1. Place 0.100 g of benzil, 0.100 g of diphenylacetone (1,3-diphenyl-2-propanone, dibenzyl ketone) and 1.0 mL of absolute ethanol in a 3 mL conical reaction vial. Place a magnetic spin vane in the vial, and attach a water-cooled jacketed condenser. Heat the mixture with stirring almost to the boiling point (look for bubbles forming on the spin vane), using the hot plate and aluminum heating block. <u>Do not boil the mixture</u>.

2. When the solids are dissolved, add 0.20 mL of ethanolic KOH (10% w/v). This solution is best added dropwise with a Pasteur pipet down through the condenser. Be sure the tip of the pipet is just above the level of solution in the vial and that the spin vane is stirring the mixture. Some frothing may occur during this addition. Add the KOH solution slowly so the frothing does not exceed the volume of your vial. As the product forms, the mixture will become a deep purple color.

3. After all of the KOH solution has been added, increase the temperature and reflux the reaction mixture gently for fifteen minutes. After the reflux period, remove the vial from the hot plate and cool to room temperature on the bench top. Then place the vial in an ice-water bath for at least five minutes to complete crystallization of the product.

4. Collect the product by vacuum filtration using a Hirsch funnel. Scrape the spin vane to remove any of the deep purple crystals that may be adhering. Use 0.8 mL of ice cold 95% ethanol to aid in the transfer of the crystals remaining in the vial to the Hirsch funnel. Wash the combined crystals again with another 0.8 mL portion of <u>ice cold</u> 95% ethanol. Allow the product to dry in your drawer in an open container until the next laboratory period.

Second Laboratory Period

1. Weigh the product and determine the melting point and percent yield. The crude product (mp 218–220°C) is of sufficient purity to be used in the next step in the synthetic sequence. The product may be recrystallized using a 1:1 mixture of 95% ethanol and toluene (25 mL of the mixed solvent will dissolve about 1 g of tetraphenylcyclopentadienone) to give a purified product with a slightly higher melting point (219–220°C).

Thin-Layer Chromatography of Substances—Prelab

Prelab Report: Due at the Beginning of the Lab Period

Name ____

Lab Section (Circle One): Mon   Tues   Wed   Thur   Fri   AM/PM

1. What is an analgesic?

2. What is an antipyretic?

3. What structural features are identical in acetaminophen, acetanilide, and phenacetin?

4. Ibuprofen is principally what kind of a drug?

5. Define R_f as applicable to TLC.

6. What are the green pigments in plants known as? What is their purpose?

7. What can you tell me about the polarity of the xanthophylls compared to carotene?

8. What are the mobile and stationary phases in the chlorophyll TLC experiment? Why are the mobile and stationary phased named that way?

9. Many kinds of intermolecular forces cause organic molecules to bind to the adsorbent on a TLC plate. Rank the strengths of these interactions (use 1 for the WEAKEST; 4 for the STRONGEST).

H-bonding   van der Waals   salt formation   dipole-dipole

__________   __________   __________   __________

PART 1. THIN-LAYER CHROMATOGRAPHY OF ANALGESICS

INTRODUCTION

In this experiment, you will use thin-layer chromatography (TLC) to determine the composition of an unknown mixture of analgesics. You will be given five known reference substances, an unknown consisting of a mixture of three of the references, as well as one unknown containing a single reference compound. The reference samples are:

Acetaminophen (AC)	
Aspirin (ASP) (also called acetylsalicylic acid)	
Caffeine (CF)	
Ibuprofen (IBU)	
Salicylamide (SAL)	

The reference samples are all available as solutions consisting of 1 g sample dissolved in 20 mL of a 1:1 mixture of methylene chloride and ethanol. The unknown mixture and the single unknowns are similarly prepared.

PROCEDURE—SPOTTING THE TLC PLATES

Working with a partner, obtain two TLC plates coated with silica gel and eight microcapillary tubes for spotting. **With pencil**, mark a baseline approximately 1 cm from the bottom of each TLC plate. Mark four lanes on each plate (see illustration below) and indicate what will be spotted on which lane. You will be spotting the five reference compounds, the unknown mixture, and two single unknowns (yours and your partner's). The lanes should be at least 0.5 cm from the side of the plate, and at least 1.25 cm apart.

Use the microcapillary tubes to spot the reference compounds, the mixture, and the two unknowns on the two plates. Be sure to use a clean microcapillary for each separate compound; please do not contaminate the reference compounds, the mixture, or the unknowns! If you are not sure if you have spotted enough compound on the plate, look at it under the UV lamp. You should see a dark purple spot on the baseline about 1-2 mm in diameter. If you can't see it, then you need to spot more compound there.

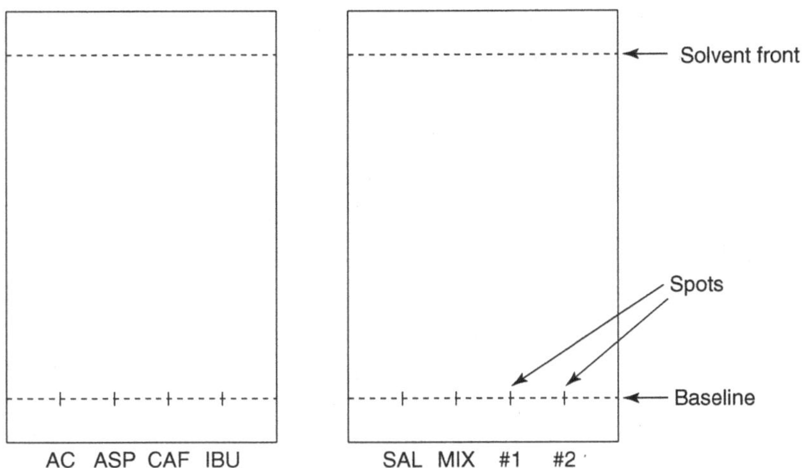

PROCEDURE—DEVELOPING THE TLC PLATES

Obtain a 16 oz. wide-mouth screw-cap jar for use as a developing chamber. Fill the chamber with the developing solvent (0.5% acetic acid in ethyl acetate) to a depth of 0.5 cm (this may already have been done for you). Note that the solvent level must not be above the baseline, or the samples will dissolve off of the TLC plate into the solvent pool rather than eluting up the plate as desired! Place both spotted TLC plates in the jar (see photograph below) and allow them to develop.

When the solvent has risen to a level about 0.5 cm from the top of the plate, remove the plates from the developing chamber and immediately draw a line in pencil on the plate at the solvent front. Put the plates on a paper towel on the bench top and allow the solvent to evaporate.

When the plate is dry, observe it under a UV lamp. Lightly outline with a pencil all of the spots that you see on the plate. Also note any particular colors or unusual fluorescence as this may assist in distinguishing compounds with similar R_f values. Trace the TLC plates and associated spots into your laboratory notebook, and note any distinguishing features.

Calculate the R_f values for each spot. Identify the components of the mixture, and identify each of the two single unknowns (be sure to record the unknown number in your notebook).

Figure 1—*TLC Jar with Two Plates*

QUESTIONS

1. Why might it be difficult to distinguish *cis-* and *trans*-2-pentene by TLC?

2. What problem will arise if the level of the developing solvent in the chamber is higher than the level at which a spot is applied to the TLC plate?

3. What would you expect to see if a TLC plate has been left in the developing chamber too long (so that the solvent front has reached the top of the TLC plate)?

4. What will the result be of spotting too much compound onto the TLC plate?

5. One of the analgesics contains a chiral center. Which one is it? One of the two enantiomers is far more effective at reducing pain than the other. Which one?

Over-The-Counter Product	Major Components
Aspirin	Aspirin! (Acetylsalicylic acid)
Excedrin	Acetaminophen, Aspirin, Caffeine
Tylenol	Acetaminophen
Advil	Ibuprofen
Motrin	Ibuprofen
Anacin	Aspirin, Caffeine
Bufferin	Aspirin
Aleve	Naproxen

PART 2. EXTRACTION AND THIN-LAYER CHROMATOGRAPHY OF CHLOROPHYLL A AND B FROM SPINACH[1]

BACKGROUND

Thin layer chromatography (TLC) is one type of chromatography. In TLC, the stationary phase is a thin layer of adsorbent particles (silica gel) attached to a solid plate. A small amount of sample is applied (spotted) near the bottom of the plate and the plate is placed in the mobile phase (70:30 hexane–acetone solvent). This solvent is drawn up by capillary action. Separation occurs as each component, being different in chemical and physical composition, interacts with the stationary and mobile phases to a different degree creating the individual bands on the plate. The retention factor, R_f value, is used to characterize and compare components of various samples.

$$R_f \text{ value} = \frac{\text{distance from origin to component spot}}{\text{distance from origin to solvent front}}$$

The pigments in vegetables, flowers, and leaves can be separated and identified by using thin-layer chromatography. Green pigments, known as chlorophylls, serve as the main photoreceptor molecules of plants. Carotenoids, yellow pigments, aid the plant in the photosynthesis process. In addition, xanthrophylls are contained in the chloroplasts which can be isolated and identified using chromatographic techniques.

Pigment	Color	R_f value in 70:30 hexane:acetone
Carotene	Yellow-orange	0.93
Pheophytin a	Grey	0.55
Pheophytin b	Light grey (may not be visible)	0.47–0.54
Chlorophyll a	Blue-green	0.46
Chlorophyll b	Green	0.42
Xanthophylls	Yellow	0.41
Xanthophylls	Yellow	0.31
Xanthophylls	Yellow	0.17

---

[1] Adapted in part by R. McLaughlin and K. Masters from Griffin, G. William; Quach, Hao T.; Steeper, Robert L. *J. Chem. Ed.* **2004**, *81*, 385–387.

Purpose

The purpose of this laboratory experiment is to isolate the pigments from spinach by using thin layer chromatography.

EQUIPMENT AND MATERIALS

TLC silica gel plates

Chromatography chambers

Test tubes

Spinach

Acetone

Sand

Capillary tubes

70:30 hexane–acetone solvent

Test tubes/metal spatula for mixing grinding

Pipettes

Anhydrous $MgSO_4$

SAFETY

- Always wear goggles and gloves in the laboratory.
- Wear gloves when handling the TLC plates.

PROCEDURE

1. Weigh out 0.5 g of fresh spinach. Combine this with 0.5 g of anhydrous magnesium sulfate and 1.0 g of sand.

2. Using a test tube and metal spatula, grind the mixture until it becomes a fine, light green powder (1–3 min).

3. Add 2 mL of acetone. Stir the solution using a stir bar for 1–2 min.

4. Cover the mixture with parafilm. Allow the mixture to sit for 5 min. The solid should settle to the bottom, leaving a green liquid layer on top.

5. Using a pipette, extract the acetone and place it in a clean small test tube. Be careful not to extract the small pieces of spinach.

6. With a pencil draw a tick mark on the edge of the TLC plate approximately 1 cm from the bottom of the TLC plate.

7. With your capillary tube, draw up some of your spinach extract and make a spot no wider than 3mm across on your TLC plate. Reapply the sample to the same place two to three times or until the spot is clearly visible. Fill the chromatography chamber to a depth of approximately 0.5 cm with the 70:30 hexane-acetone mobile phase. (Your mobile phase SHOULD NOT be higher than your TLC spot.) Place the TLC plate in the chromatography chamber with the sample spot toward the bottom. Close the chamber. Make sure not to shake the chamber as the mobile phase with slosh and rinse your spot down the plate. Should this happen, dry the plate and re-spot.

8. Allow the plate to remain undisturbed until the solvent reaches to within 1 cm of the top.

9. Remove the plate from the chamber and *immediately mark the solvent front* using a pencil.

10. Measure and record the distance from the spotting line (origin) to the center of each spot and from the spotting line to the solvent front.

11. Lightly circle the individual bands with a pencil after the plate dries because the pigment will eventually fade. IMMEDIATELY RECORD YOUR COLOR RESULTS.

12. Identify each component (spot).

Record this table in your notebook, along with a sketch/diagram of your TLC plate, followed by the calculations and questions.

Data Table

Distance solvent moved from the spotting line (origin) ______________________________

Color of spot	Distance moved	R_f value	Identity

Calculations

Calculate the R_f value for each spot observed.

QUESTIONS

1. Explain the color of the spinach based on the results of the chromatography.

2. Why was $MgSO_4$ used? Sand?

3. Why do you think TLC (or chromatography in general) is a really important analysis to use when a new plant or shrub/tree is discovered?

THE WITTIG REACTION—PRELAB

Prelab Report: Due at the Beginning of the Lab Period

Name ____

Lab Section (Circle One): Mon   Tues   Wed   Thur   Fri   AM/PM

1. What is an ylide?

2. What color do you expect your ylide to be?

3. Show the complete mechanism for this reaction, beginning with triphenylphosphine, benzyl chloride, sodium ethoxide, and cinnamaldehyde.

4. What solvent will you use to develop the TLC plate to analyze your reaction?

5. What spot do you predict will have the *lowest* R_f value?

6. If you started with 0.354 g benzyltriphenylphosphonium chloride and 0.100 mL of cinnamaldehyde, and obtained a 76% yield, what is the weight <u>in grams</u> of your isolated product? (Show your calculations.)

THE WITTIG REACTION

BACKGROUND

The Wittig reaction is commonly used to convert carbonyl compounds into alkenes. In this experiment, cinnamaldehyde will react with a Wittig reagent, benzyltriphenylphosphonium chloride, to produce two isomeric dienes: *cis,trans-* and *trans,trans-*1,4-diphenyl-1,3-butadiene. Only the *trans,trans-* isomer will actually be isolated.

The reaction has three fundamental steps:

1. First, the phosphonium salt is formed by an S_N2 displacement of a 1° or 2° alkyl halide with triphenylphosphine. The product is referred to as the Wittig reagent or Wittig salt. (In this particular experiment, benzyl chloride would react with triphenylphosphine to produce benzyltriphenylphosphonium chloride; this Wittig reagent is commercially available, and we will not actually make it in the laboratory.)

2. Next, the Wittig salt is reacted with a strong base. This could be an alkyl lithium reagent or another base such as sodium ethoxide. The product is called an ylide (a species having adjacent atoms with opposite charges). Phosphorus ylides are quite stable, as phosphorus is not a first-row element and therefore is not limited by the octet rule. Phosphorus can use its $3d$ orbitals to overlap with the $2p$ orbital on the adjacent carbon, giving a high level of resonance stabilization and thereby stabilizing the resultant carbanion:

The ylide is an excellent source of nucleophilic carbon, and when a compound containing a carbonyl group such as an aldehyde or ketone is added to the ylide, it adds to the carbonyl group.

3. The ylide and carbonyl compound react in a [2+2] fashion to form a four-membered-ring intermediate called an oxaphosphetane. This intermediate is quite unstable and quickly fragments into an alkene and triphenylphosphine oxide. The cyclic intermediate does not break open to form the same compounds from which it was formed; instead, it breaks the opposite way. The driving force for this ring-opening process is the formation of the very stable triphenylphosphine oxide:

SPECIAL INSTRUCTIONS

1. The prepared sodium ethoxide solution must be kept tightly covered when not being used, as it reacts readily with atmospheric water. Freshly distilled cinnamaldehyde must be used, as old cinnamaldehyde tends to contain significant amounts of cinnamic acid. The cinnamaldehyde will be checked by IR to be sure it is pure.

2. Be sure to dispose of all waste materials properly. Check with your laboratory instructor if you are unsure of the proper disposal method.

PROCEDURE

Be sure to dry the 5-mL reaction vial in the oven for at least twenty minutes. Water is anathema in this experiment.

Preparation of the Ylide

1. Place 0.480 g of benzyltriphenylphosphonium chloride in a dry, 5-mL reaction vial containing a magnetic spin vane. Add 2 mL of absolute (anhydrous) ethanol to the vial, and stir to dissolve.

2. Add 0.75 mL of the sodium ethoxide/ethanol solution to the vial while stirring on the magnetic stirrer. Cap the vial, and continue stirring for fifteen minutes. During this time, the solution will become cloudy, and you will see the characteristic yellow color of the ylide. (Ylides in general tend to be highly colored: sometimes pink, sometimes purple, sometimes yellow.)

Reaction of the Ylide with Cinnamaldehyde

Obtain 0.15 mL of pure cinnamaldehyde, and place it in another reaction vial with 0.50 mL of absolute ethanol. Cap the vial and swirl to be sure the cinnamaldehyde is dissolved in the ethanol. After the fifteen-minute reaction period to form the ylide, add the cinnamaldehyde solution to the ylide with a Pasteur pipet. A color change should be observed as the two components react. Continue stirring for another ten minutes.

Separation of the Alkene Isomers

1. Cool the vial thoroughly (ten minutes) in an ice-water bath, stir with a small metal spatula, and collect the solid by vacuum filtration using a Hirsch funnel. Rinse the vial and product with two 1 mL portions of ice-cold absolute ethanol. At this point, the Hirsch funnel contains the solid *trans,trans* alkene as well as NaCl (see the reaction scheme above if you're not sure where the NaCl came from). The cloudy liquid in the filter flask contains triphenylphosphine oxide, the more soluble *cis,trans* alkene, and a small amount of the *trans,trans* alkene. At this point, pour the filtrate into a labeled beaker, and save it for further analysis as described below.

2. To remove the NaCl from the product, scrape the solid from the Hirsch funnel, and place it in a small beaker. Add 3 mL of cold water, stir thoroughly, and refilter. Use another 1 mL of cold water, if necessary, to aid in the transfer.

Analysis of the Filtrate

1. You will analyze your product using TLC. The analysis must be done quickly, as the *cis,trans* alkene isomer will photochemically convert to the more stable *trans,trans* isomer if it sits out for too long. Spot your filtrate on one lane of the TLC plate. Dissolve a small amount of the solid *trans,trans* alkene in a few drops of acetone, and spot that solution on another lane. (You should share a TLC plate with another student, as there is room for four lanes on one plate.) Develop the TLC plate using ligroin as the solvent.

2. Visualize the spots on the TLC plate with a UV lamp, using both short and long wavelength settings. Identify the triphenylphosphine oxide, the *cis,trans* alkene, and the *trans,trans* alkene (this one fluoresces brightly). Be sure to trace a copy of the TLC plate in your laboratory notebook, calculate all R_f values, and identify all spots.

Yield and Melting Point

When the *trans,trans*-1,4-diphenyl-1,3-butadiene is dry, determine the melting point and weight of your sample. Calculate the percent yield. If the melting point is below 145°C, recrystallize your sample from 95% ethanol (approximate solubility 20 mg/1.3 mL ethanol), and redetermine the melting point.

QUESTIONS

1. What conclusions can you draw about the contents of the filtrate and the purity of the *trans,trans* product from your TLC results?

2. There is an additional isomer of 1,4-diphenyl-1,3-butadiene, which has not been shown in this experiment. Draw the structure and name it. Why is it not produced in this experiment?

3. Why should the *trans,trans* alkene be the thermodynamically most stable isomer?

PART IV

APPENDICES

APPENDIX I

MICROKIT GLASSWARE

3- and 5-mL reaction vials

Teflon spin vane

10-mL rb flask

Illustrations created by Marietta H. Schwartz

Air condenser and jacketed condenser

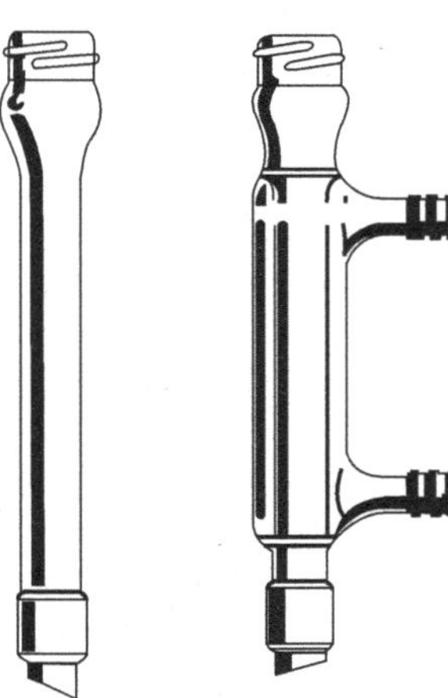

Drying tube

Crag tube w/plunger

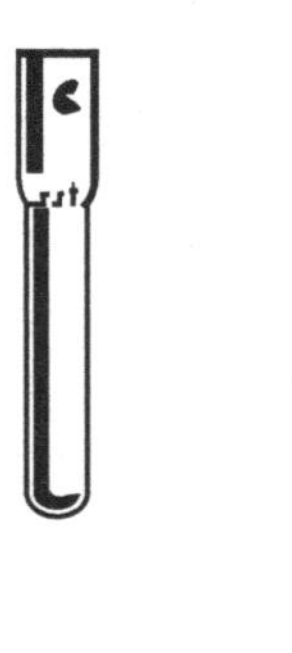

Illustrations created by Marietta H. Schwartz

Thermometer adapter

Vacuum adapter

Claisen head

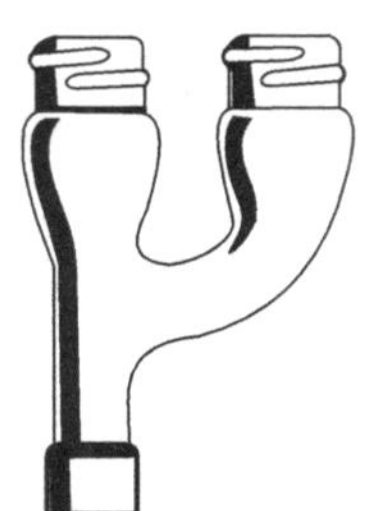

Distilling head

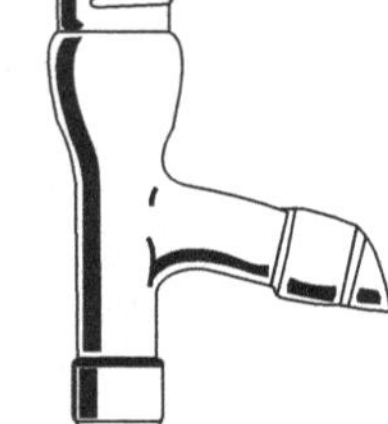

Illustrations created by Marietta H. Schwartz

OTHER GLASSWARE AND EQUIPMENT

Beakers   Erlenmeyer flasks   Graduated cylinders   Centrifuge tubes

 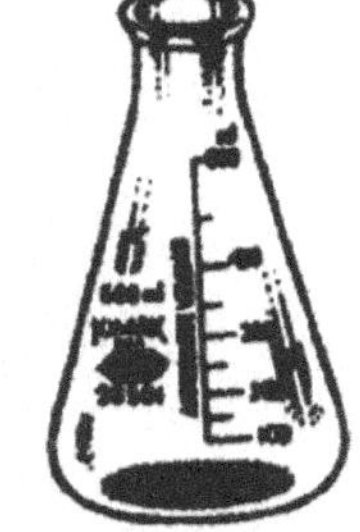  

Illustrations created by Marietta H. Schwartz

Watch glasses   Thermometers   Spatula/Scoopula   Wash bottle

Forceps (tongs)

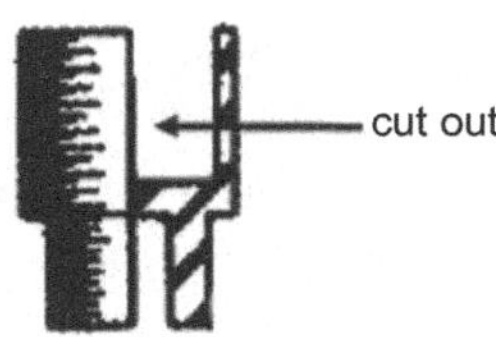  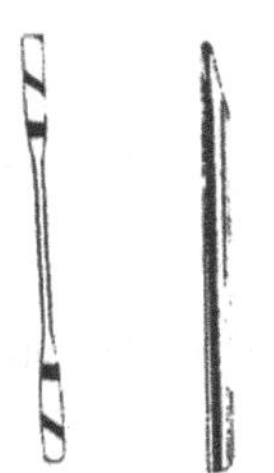 

Illustrations created by Marietta H. Schwartz

Rubber septum   Hirsch funnel w/frit (0) and filter flask (N)   Powder funnel

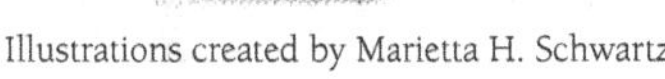 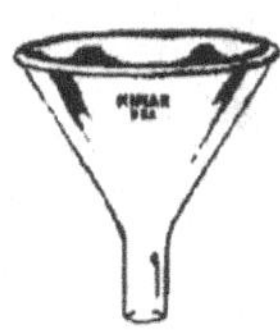

Illustrations created by Marietta H. Schwartz

APPENDIX **II**

CONCENTRATED ACIDS AND BASES

Reagent	HCl	HNO_3	H_2SO_4	HCO_2H	CH_3CO_2H	$NH_3(NH_4OH)$
Density (g/mL)	1.18	1.41	1.84	1.20	1.06	0.90
% Acid or Base (by weight)	37.3	70.0	96.5	90.0	99.7	29.0
Molecular Weight (g/mol)	36.47	63.02	98.08	46.03	60.05	17.03
Molarity of Concentrated Acid or Base	12	16	18	23.4	17.5	15.3
Normality of Concentrated Acid or Base	12	16	18	23.4	17.5	15.3
Volume of Concentrated Reagent Required to Prepare One Liter of 1 M Solution (mL)	83	64	56	42	58	65
Volume of Concentrated Reagent Required to Prepare One Liter of 10% Solution by Weight (mL)	227	101	56	93	58	384
Molarity of a 10% Solution (by Weight)	2.74	1.59	1.02	2.17	1.67	5.87

APPENDIX **III**

COMMON ORGANIC SOLVENTS

Solvent	Boiling Point (°C)	Density (g/mL)
Acetone	56	0.79
Benzene	80	0.88
1-Butanol	118	0.81
Chloroform	61	1.48
Ethanol	78	0.80
Diethyl Ether	35	0.71
Ethyl Acetate	77	0.90
Hexane	69	0.66
Ligroin	60–90	0.68
Methanol	65	0.79
Methylene Chloride	40	1.32
Petroleum Ether	30–60	0.63
1-Propanol	98	0.80
2-Propanol (isopropanol)	82	0.79
Tetrahydrofuran	65	0.99
m-Xylene	139	0.87

APPENDIX **IV**

PROBLEM-SOLVING HINTS FOR SPECTRAL INTERPRETATION

1. Using the molecular formula calculate SODAR (<u>S</u>um <u>O</u>f <u>D</u>ouble bonds <u>A</u>nd <u>R</u>ings).

    Formula: SODAR = [2(#C's) +2 + (#trivalents) − (#monovalents) − (#H's)]/2

    This should always be an *integer* ⩾ 0.

2. If given the IR spectrum, check for the presence/absence of functional groups (alcohol, carbonyl, unsaturation, etc.).

3. If given the ^{1}H NMR spectrum, analyze as follows:

    a. Look at the *integrals* to calculate the number of protons responsible for each peak.

    b. Look at the *splitting patterns* to get information about the number of neighboring hydrogens, and to establish pieces of the connectivity of the molecule.

    c. Look at the *chemical shift* of each peak to get information about the location of electronegative elements contained within the molecule.

4. Once you have assembled the various pieces of the molecule from the information given, put them together so as to achieve a "reasonable" solution. This means: no pentavalent carbons, no trivalent carbons, the connectivity should match the observed splitting patterns, the functional groups should match the IR, the number of atoms should match the molecular formula . . .

5. Be sure to double-check your final answer with the spectral data—does it make sense?

6. SHOW YOUR WORK!!

Characteristic Infrared Ranges

Functional Group		Range (cm^{-1})
Alkane	(C–H)	2800–3000
	(–CH$_3$)	1360–1380
Alkene	(=C–H)	3050–3150
	(=CH$_2$)	890 (strong)
	(H–C=C–H)	*trans* 970 (strong)
	(C=C)	1620–1680
Alkyne	(C≡C–H)	3300
	(C≡C)	2100–2260
Alcohol	(O–H)	3200–3500 (broad)
Ether	(C–O–C)	1050–1280
Aldehyde	(O=C–H)	2700–2850
	(C=O)	1710–1735
	(Aromatic C=O)	1690–1700
Ketone	(C=O)	1700–1720
	(Aromatic C=O)	1660–1670
Carboxylic Acid	(O–H)	2500–3200 (broad)
	(C=O)	1700–1730
Aromatic	(=C–H)	3050–3150
		Also 1600, 1500, 900–690
Ester	(C–O–C)	1050–1280 (2 bands)
	(C=O)	1720–1740
Amine	(N–H)	3300–3500 (2 bands for 1° amine, 70 cm^{-1} apart; 1 band for 2° amine; absent for 3° amine)
Amide	(N–H)	3100–3500; also 1550–1640 for 1° and 2°
	(C=O)	1650–1670
Nitrile	(C≡N)	2220–2260

Characteristic ^{1}H NMR Chemical Shifts

Functional Group	Proton Type	Range (ppm)
Alkane	CH_3	0.9
	CH_2	1.3
	CH	1.7
Carbonyl	$O=C–CH_3$	2.1–2.5
Alkyne	$C≡C–H$	2.4
Halogen	$X–C–H$	2–4 (varies with electronegativity of halogen)
Alkoxy	$O–CH_3$	3.5–4.1
Alkene	$=C–H$	4.5–5.5 (moves downfield with conjugation)
Aromatic	$Ph–H$	7–8
Aldehyde	$O=C–H$	9.7
Carboxylic Acid	$O=C–O–H$	10.5–12
Alcohol	$O–H$	0.5–4 (quite variable; usually broad singlet)
Amine	$N–H$	0.5–5 (quite variable; usually broad singlet)

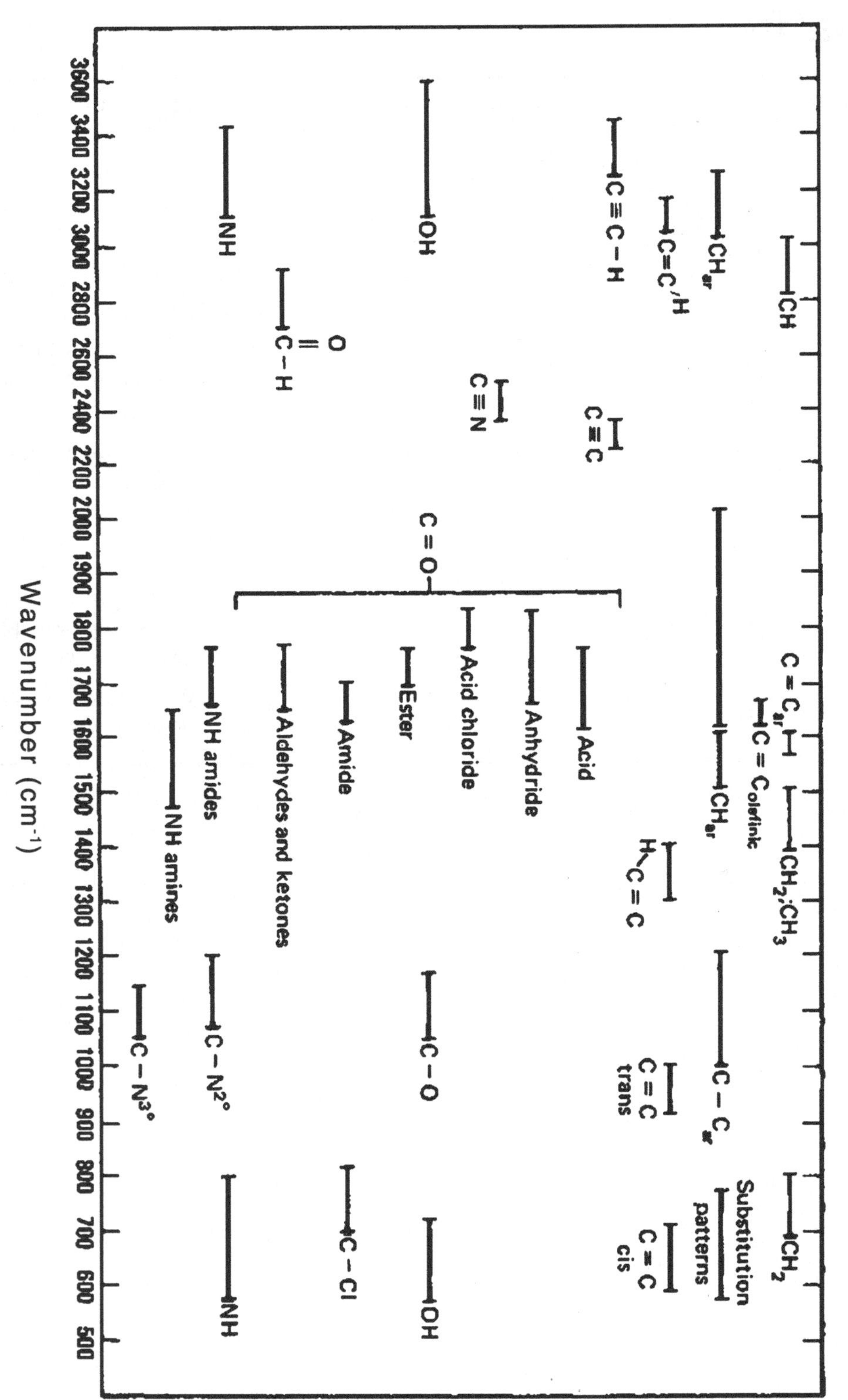

Adapted from http://what-when-how.com/organic-chemistry-laboratory-survival-manual/infrared-spectroscopy-part-1-laboratory-manual/

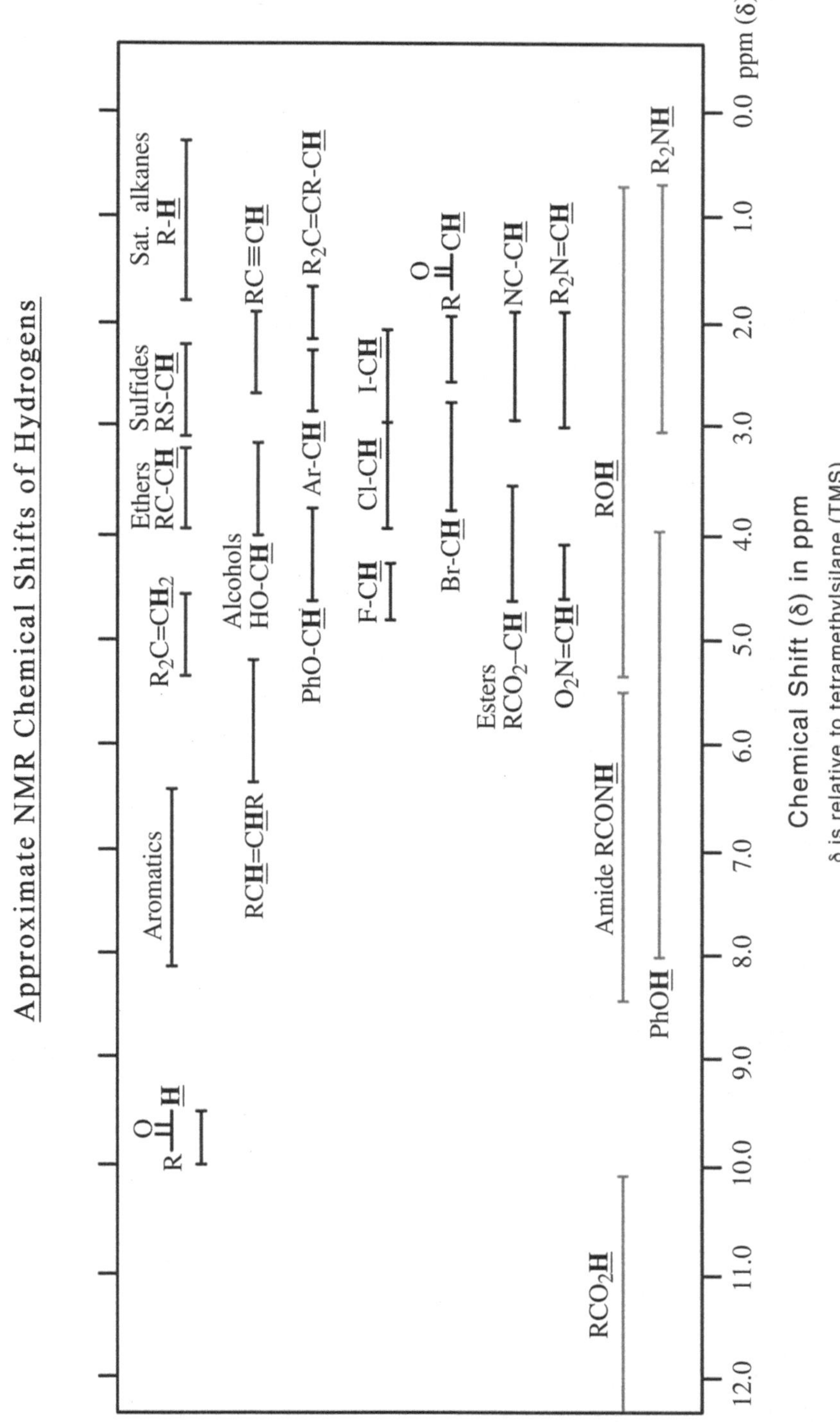

Adapted from https://www2.chemistry.msu.edu/faculty/reusch/virttxtjml/spectrpy/nmr/nmr1.htm

INTERPRETATION OF INFRARED SPECTRA

HYDROCARBONS

Hydrocarbons show IR absorption peaks between 2800 and 3300 cm^{-1} due to C–H stretching vibrations. The hybridization of the carbon affects the exact position of the absorption—stiffer bonds vibrate at higher frequencies. sp^3 C–H: 2800–3000, sp^2 C–H: 3000–3100, sp C–H: 3300 cm^{-1}.

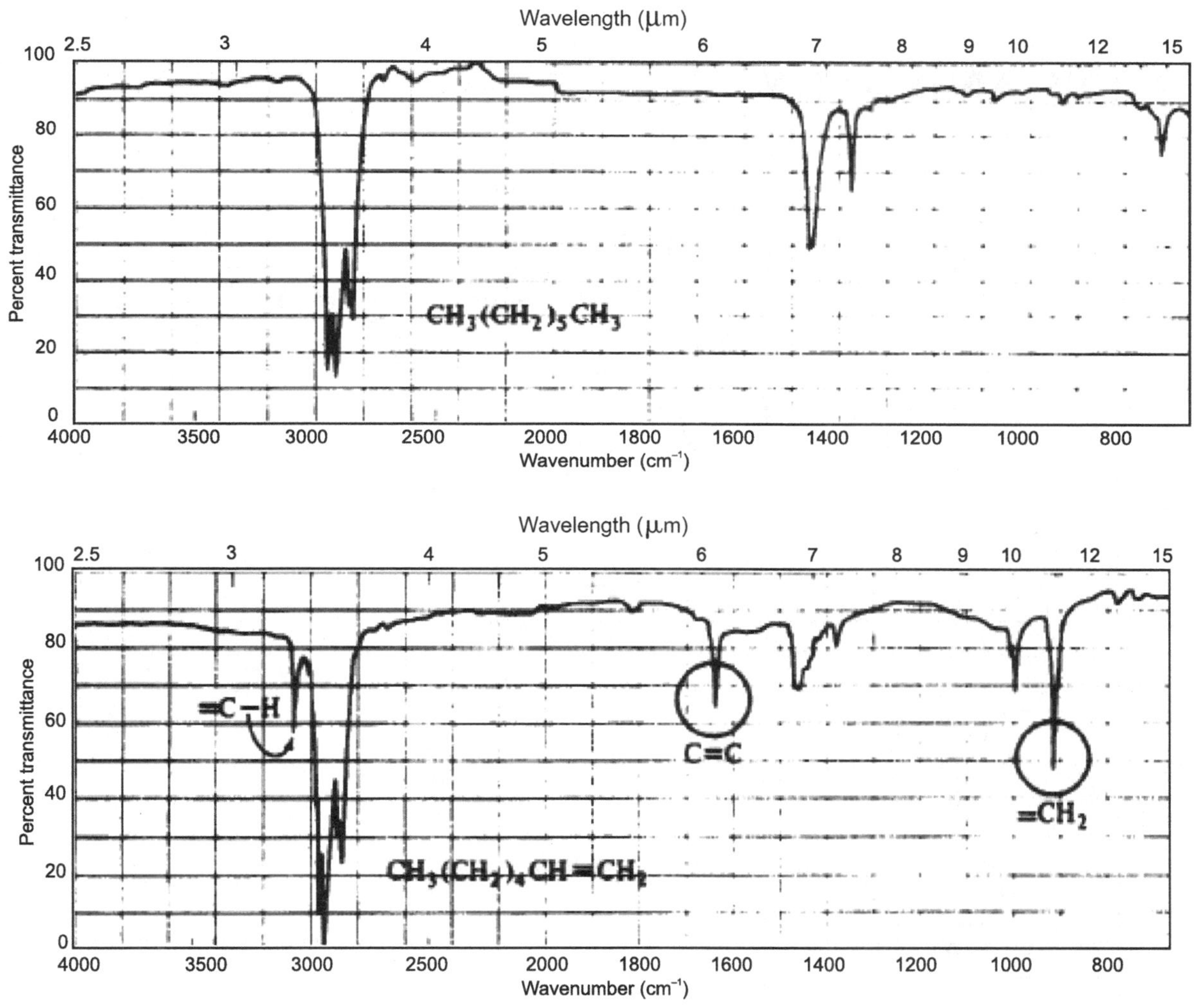

From Fessenden/Fessenden. *Organic Chemistry*, 3E. © 1986 Brooks/Cole, a part of Cengage, Inc. Reproduced by permission. www.cengage.com/permissions

AROMATIC COMPOUNDS

The position of substitution on a benzene ring can sometimes be determined from the IR spectrum. Benzene rings often give characteristic absorptions at about 680–900 cm^{-1}. The patterns observed are summarized in the following table:

Substitution Pattern	Appearance	Position of Absorption (cm^{-1})
monosubstituted	two peaks	730–770
		690–710
o-disubstituted (1,2)	one peak[a]	735–770
m-disubstituted (1,3)	three peaks	860–900
		750–810
		680–725
p-disubstituted (1,4)	one peak	800–860

[a] An additional peak is often observed at approximately 680 cm^{-1}.

NITRILES AND ALKYNES

The CN triple bond absorption appears at 2200–2300 cm^{-1}, in about the same place as the CC triple bond absorption. Both of these bands are usually medium to weak in intensity.

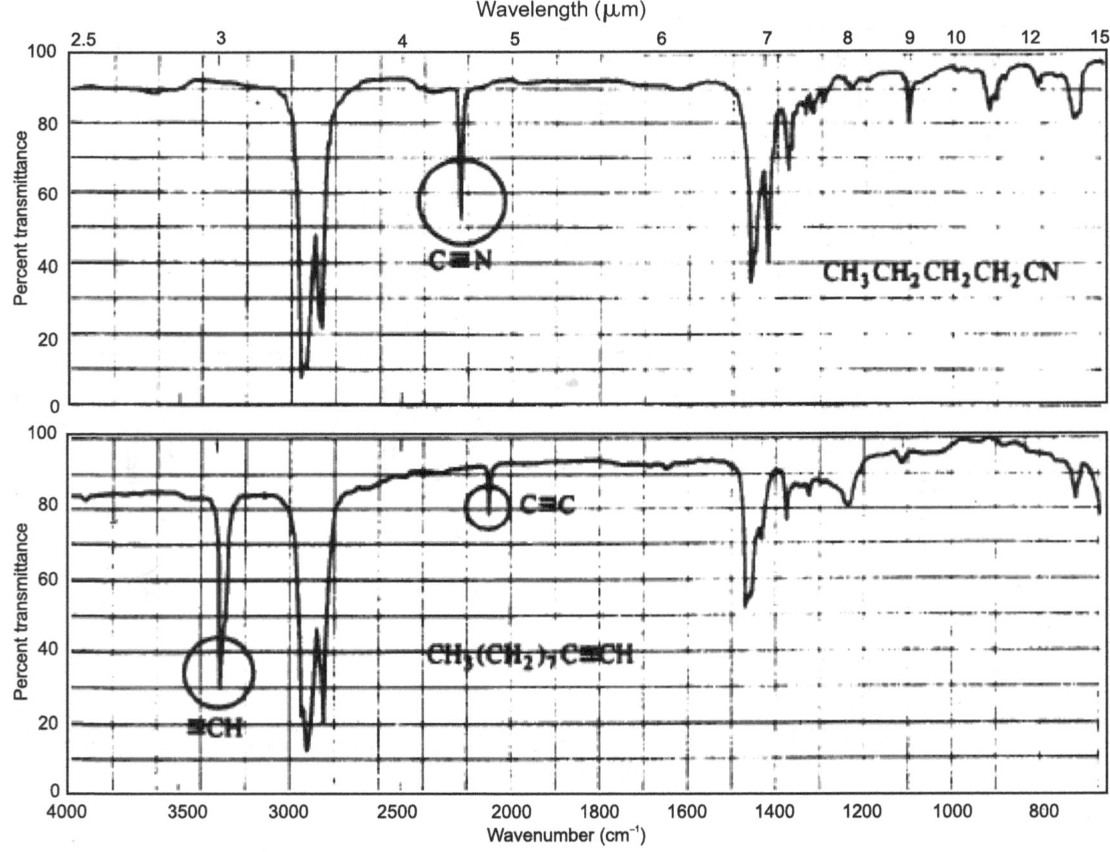

ALCOHOLS AND AMINES

Alcohols and amines show a conspicuous OH or NH stretching absorption at 3000–3700 cm^{-1} (to the left of the hydrocarbon CH stretch). If there are two hydrogens on the amine, a double peak is seen. If the compound is a tertiary amine, no NH stretch is observed. The OH absorptions are generally quite intense and smoothly curved. NH stretches are weaker and narrower.

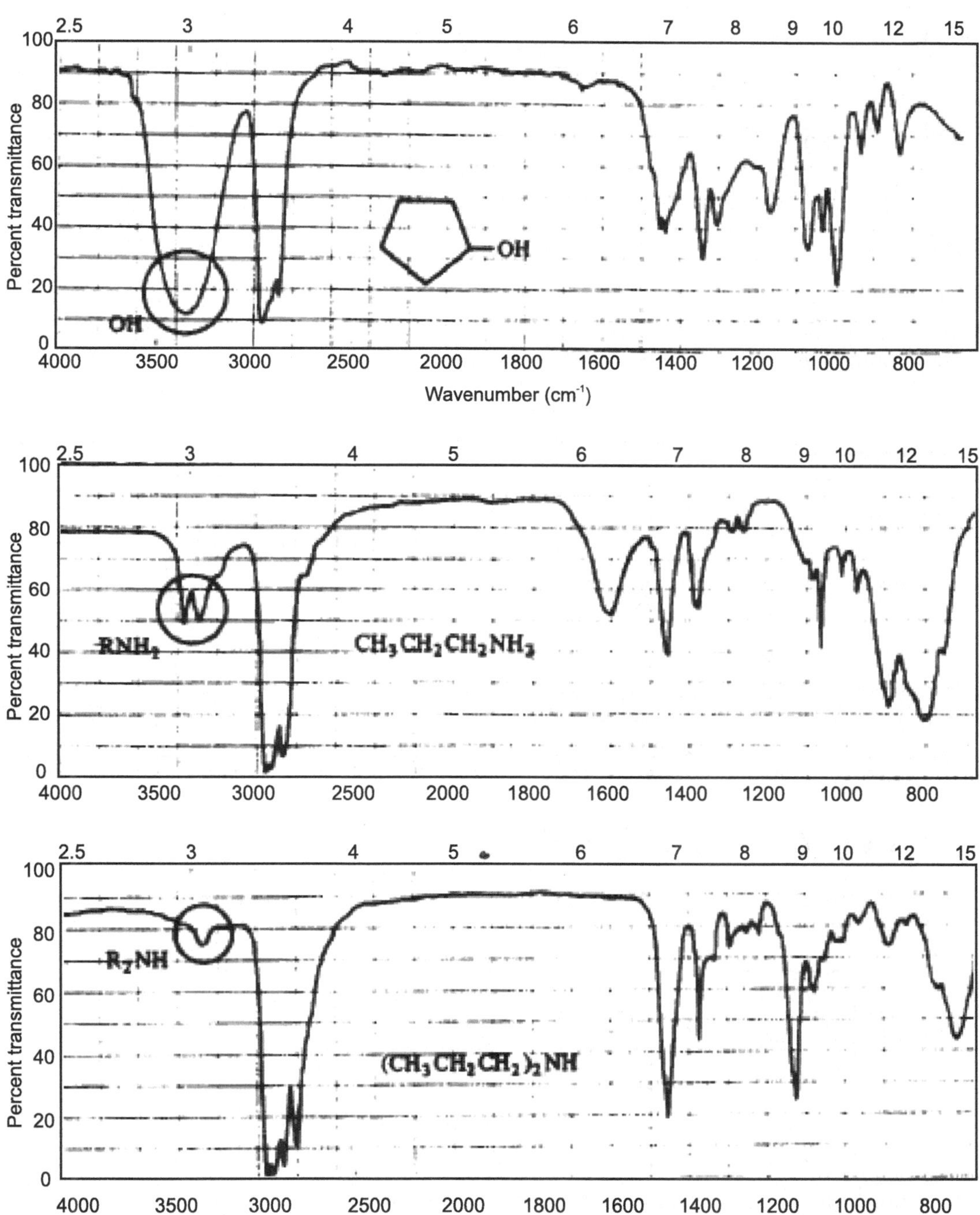

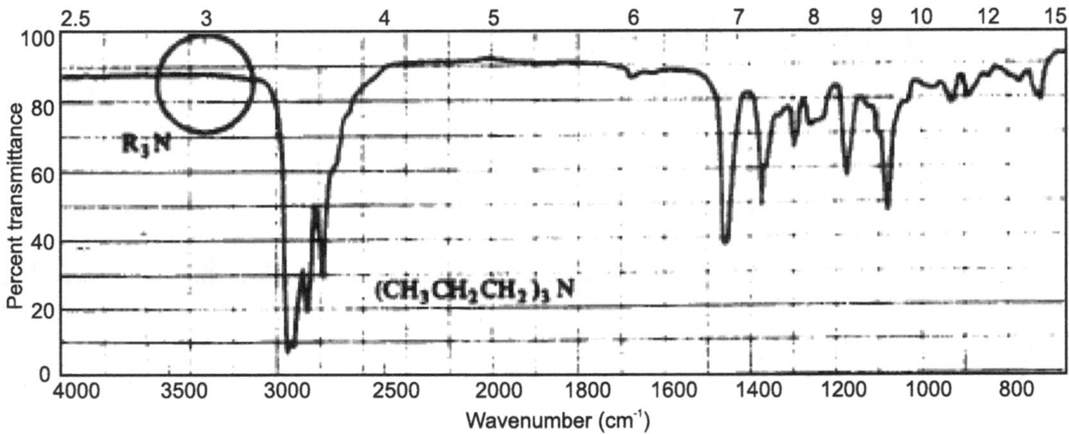

ETHERS

Ethers have a C–O stretch that appears in the fingerprint region at 1050–1260 cm⁻¹. This is generally a strong absorption, but can be difficult to detect if the fingerprint region is complex. Alcohols, esters, and other compounds containing C–O single bonds also show a C–O stretch in this region.

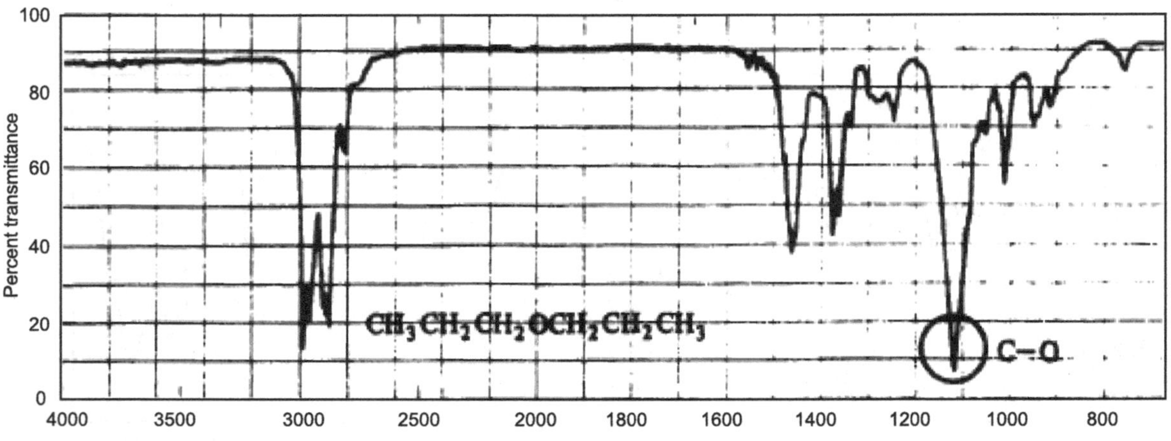

CARBONYL COMPOUNDS

The carbonyl (C=O) stretch is one of the most distinctive bands in the IR spectrum. It can be found at 1640–1820 cm⁻¹ and is generally a very strong peak.

Functional Group	Position of Absorption (cm⁻¹)
Aldehyde (RCHO)	1720–1740
Ketone (RCOR')	1705–1750
Carboxylic Acid (RCO₂H)	1700–1725 (also note broad OH stretch)
Ester (RCO₂R')	1735–1750 (also note C–O stretch)

It should be pointed out that the C=O absorption ranges for the various carbonyl-containing functional groups overlap significantly, so it is difficult to make a definitive identification of the functional group based solely on the position of the carbonyl peak. However, most of these functional groups show other diagnostic absorptions that assist in identification. These are summarized below, and illustrated on the following pages.

Aldehydes: will also show a distinctive C–H stretch around 2800 cm^{-1}.

Ketones: will *not* show any of the other distinctive absorptions mentioned here.

Carboxylic Acids: will also show a very broad OH stretch that frequently obscures the CH stretch around 3000 cm^{-1}.

Esters: will also show a strong C–O single bond stretch as discussed earlier.

Carboxylic Acids

Have a very distinctive OH band which usually starts around 3300 cm^{-1} and stretches into the aliphatic CH region, often partially or completely obscuring the CH stretch.

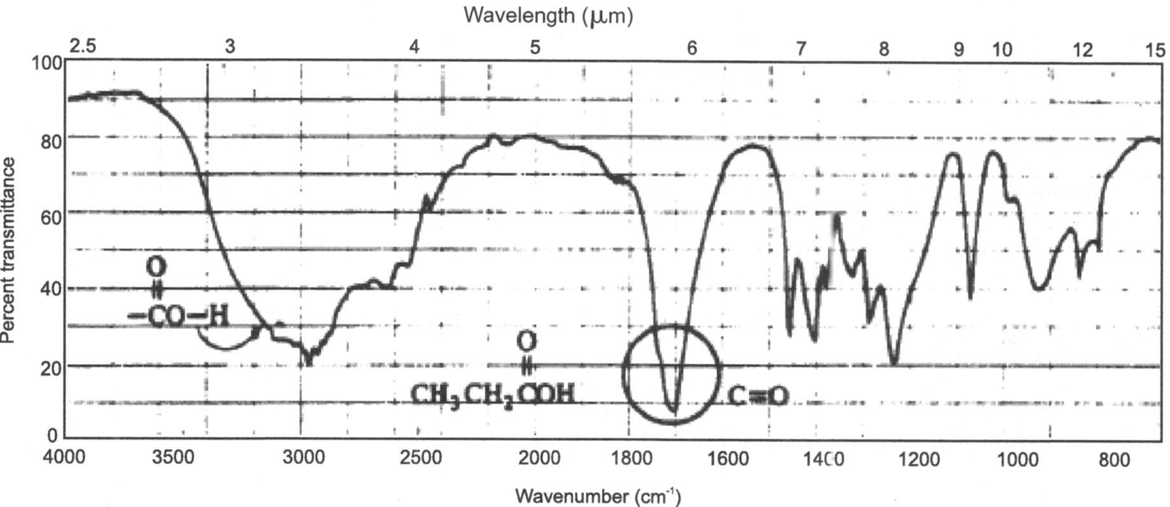

Esters

Have both a C=O and a C–O stretch.

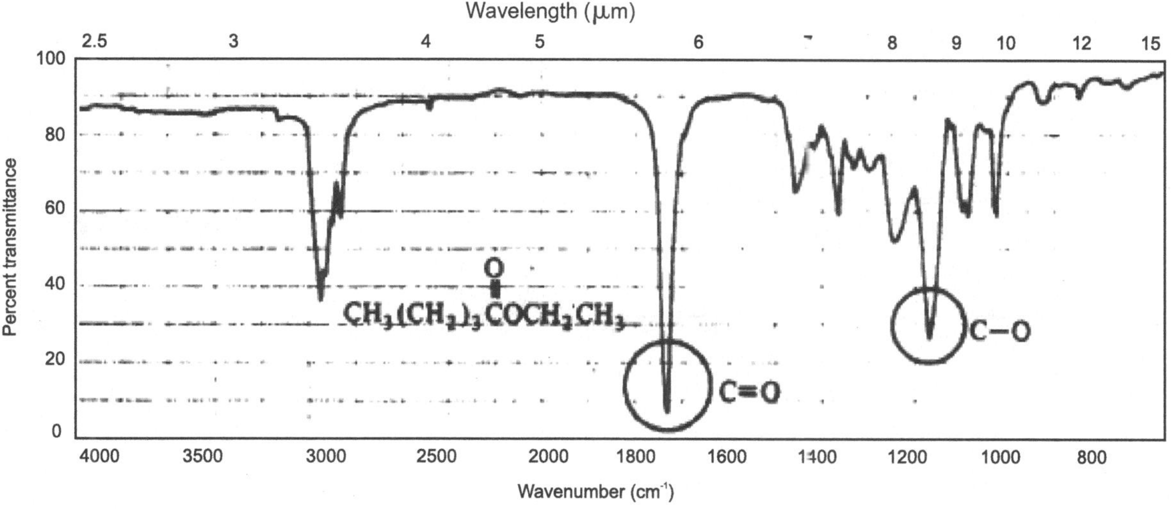

KETONES

Have the simplest spectra of the carbonyl compounds, with the C=O stretch and the C–H stretch being the major absorptions (unless other functionality is present).

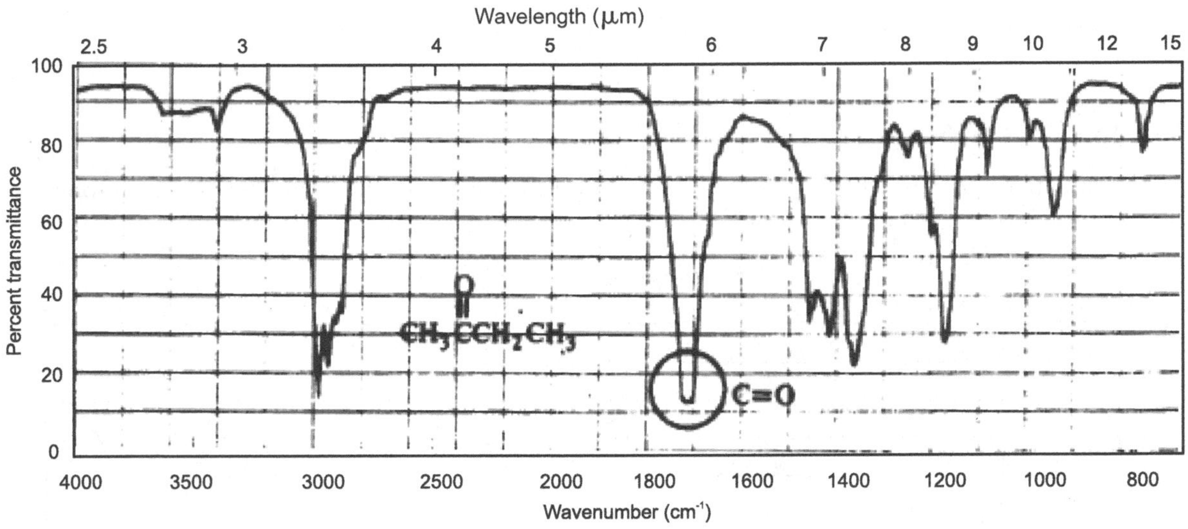

ALDEHYDES

The important difference between aldehydes and ketones is the aldehydic CH stretch. This absorption usually appears to the right of the sp3 CH stretch, at 2700–2900 cm⁻¹.

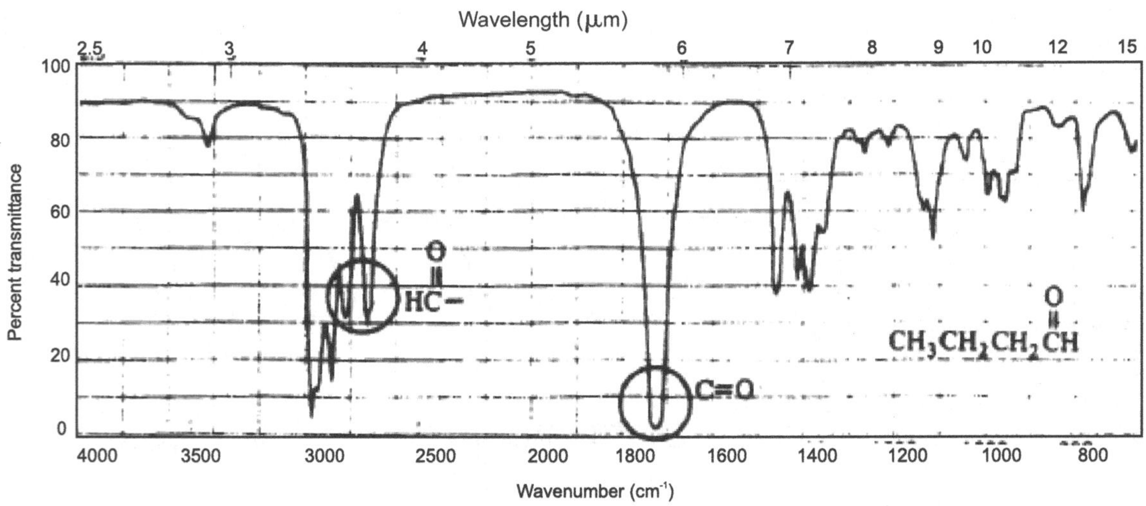